The Ilímaussaq Alkaline Complex, South Greenland

– an Overview of 200 years of Research and an Outlook

Contribution to the Mineralogy of Ilímaussaq no. 130

Henning Sørensen

MEDDELELSER OM GRØNLAND • GEOSCIENCE 45 • 2006

Henning Sørensen: The Ilímaussaq Alkaline Complex, South Greenland – an Overview of 200 years of Research and an Outlook. Contribution to the Mineralogy of Ilímaussaq no. 130. – Meddelelser om Grønland Geoscience 45. Copenhagen, The Commission for Scientific Research in Greenland, 2006.

Editor in publication: Kirsten Caning
Printed by Special-Trykkeriet Viborg a-s

Front cover: Sketch of the west contact (at the extreme right) of the Ilímaussaq complex in Nunasarnaasaq mountain (767 m) at the head of Kangerluarsuk. The nepheline syenites are separated from the basement granite overlain by sandstone and basaltic sheets by a rim of augite syenite (the escarpment). Water colour made in 1876 by Andreas Kornerup (1857-1881).
Back of cover: (Top) The south wall of the Sermilik fjord north of Tunulliarfik showing in the lower part Gardar sandstone with basaltic sheets and in upper part basaltic lavas. The highest peak is Ilimmaasaq mountain (1390 m). (Bottom) The south side of Kangerluarsuk showing from right to left: The basement granite in the Iviangiusaq mountain (903 m), (cf. Fig. 21), then the augite syenite rim (light colour) and after that the layered kakortokite and at the left Laksefjeld (680 m) showing transition from kakortokite to lujavrite (cf. Fig. 24). In the background, Killavat ridge (1216 m). Water colour by Andreas Kornerup 1876

Scientific Editor: Dr. Svend Funder, University of Copenhagen, Geological Museum, Øster Voldgade 5-7, DK 1350 Copenhagen K. Phone + 45 3532 2345, fax + 45 3332 2325, svf@savik.geomus.ku.dk

About the monographic series *Meddelelser om Grønland*
Meddelelser om Grønland is Danish for *Monographs on Greenland*. *Meddelelser om Grønland* has been published by the Commission for Scientific Research in Greenland, Denmark, since 1879. Since 1979 each publication is assigned to one of the three subseries: *Man & Society, Geoscience, Bioscience.*
Geoscience invites papers that contribute significantly to studies in Greenland within any of the fields of geoscience (physical geography, oceanography, glaciology, general geology, sedimentology, mineralogy, petrology, palaeontology, stratigraphy, tectonics, geophysics, geochemistry).
For more information and a list publications, please visit the web site of the Danish Polar Center *http://www.dpc.dk/PolarPubs*

All correspondence concerning this book or the series *Meddelelser om Grønland* should be sent to:
Danish Polar Center, Strandgade 102, DK-1401 Copenhagen, Denmark
tel +45 3288 0100, fax +45 3288 0101, email dpc@dpc.dk, www.dpc.dk

Accepted May 2006
ISSN 0106-1046
ISBN 87-90369-85-8

Contents

Abstract

Henning Sørensen: The Ilímaussaq Alkaline Complex, South Greenland – an Overview of 200 years of Research and an Outlook. Contribution to the Mineralogy of Ilímaussaq no. 130. – Meddelelser om Grønland Geoscience 45. Copenhagen, The Commission for Scientific Research in Greenland, 2006. 70 pp.

The Ilímaussaq alkaline complex, South Greenland, the type locality of agpaitic nepheline syenites and of thirty minerals, among which arfvedsonite, eudialyte and sodalite, has been studied since 1806. The paper reviews the outcome of 200 years of geological investigations and presents an overview and a synthesis of the petrology of the complex. The site of the complex was invaded successively by augite syenitic, alkali granitic and nepheline syenitic melts. Remnants of the augite syenite and granite intrusions are found along the contacts of the nepheline syenites and as xenoliths in these. The nepheline syenites, which occupy the major part of the complex, are divided into a roof series, an intermediate series and a floor series. The roof series crystallized from the roof downward in the order pulaskite, foyaite, and the agpaitic rocks sodalite foyaite and naujaite. The floor series consists of cumulates: an inferred hidden part formed simultaneously with the roof series and an exposed part of layered agpaitic nepheline syenites (kakortokites) which were formed at least partly later than the roof series. The floor series passes gradually into the overlying intermediate series consisting of the agpaitic rock type called lujavrites. The lujavrites enclose rafts of naujaite and appear to have been emplaced by piecemeal stoping. The larger lujavrite masses are floor cumulates, but lujavrite dykes and sheets occupy fractures in the roof series rocks.

According to one model for the evolution of the complex, the nepheline syenites formed by consolidation of one magma batch in a closed system. The lujavrites were formed from the residual melts left after the formation of the roof series and the floor series and were sandwiched between these. A second model implies that the kakortokites and lujavrites formed from one or more separate magma pulses which intruded the already consolidated roof series rocks. This model is supported by new information on contact relations and especially on the petrology and geochemistry of a marginal pegmatitic facies that forms a rim around the kakortokite-lower aegirine lujavrite part of the complex. It consists of a massive-textured matrix intersected by pegmatites. The matrix was the first rock to form in the lowermost exposed part of the complex and gives information about the composition of the initial magma of the kakortokite-lujavrite sequence.

It is concluded that the agpaitic rocks of the complex were formed from at least two successive magma injections, which formed respectively the roof series and the kakortokite-lujavrite sequence.

Key-words: agpaitic, Ilímaussaq, Gardar, Greenland, lujavrite, naujaite

Henning Sørensen, Geological Institute, University of Copenhagen, Øster Voldgade 10, DK-1350 Copenhagen K. fax: (45) 35322440, e-mail: hennings@geol.ku.dk.

Introduction

The Ilímaussaq alkaline complex, South Greenland, has been recognised as a unique geological area since K.L. Giesecke undertook the first examination of its mineral wealth in 1806. The complex is the type locality of agpaitic nepheline syenites (Ussing 1912) and of thirty minerals, among them arfvedsonite, eudialyte and sodalite (Petersen 2001). Twelve minerals are unique to this place. More than 700 papers have been published on the geology, mineralogy, petrology, geochemistry and economic geology of the complex (Rose-Hansen *et al.* 2001). The reprint series *Contributions to the Mineralogy of Ilímaussaq,* which was initiated in 1965, has reached no.130 in 2006; nos. 63 and 100 of the series are collections of papers. A review of the history of investigation of the complex with a brief introduction to its geology is presented by Sørensen (2001).

The author of this paper first visited the complex in 1946 and began regular work on it in 1955. During the more than 50 years that have passed since then, a considerable amount of knowledge has been accumulated by a multinational group of researchers. But it has also become increasingly clear that in spite of the great number of man-years spent on the complex, much remains to study and explain as it will be demonstrated in the present paper.

Most of the agpaitic rocks of the complex are cumulates and some of them are extremely coarse-grained. This makes sampling of rocks whose chemical composition may be regarded to be equivalent to liquid compositions difficult. The paper describes some hitherto ignored rocks which nevertheless may represent melt compositions and thus provide new information about agpaitic melts. They form a rim, the so called marginal pegmatite, along the contact between parts of the agpaitic rocks of the complex and their country rocks. This rim is made up of agpaitic nepheline syenites which are intersected by short pegmatite veins. In some places, the matrix consists of unlayered, massive-textured agpaitic rocks. Massive-textured fine-grained rocks also form pockets in pegmatites and are considered to have been formed by quenching of melts in response to a release of volatiles. Thus, the massive-textured rocks may give information about the composition of the melts from which they were formed. A number of such rocks have been chemically analysed and are treated in some detail in the present paper.

The construction of the agpaitic part of the complex is occasionally mimicked on a small scale in pegmatites; some examples are described and discussed.

The present paper highlights 200 years of reseach in the complex by bringing an overview of the main features of its geology based on the extensive literature, the author's own experience and on hitherto unpublished observations of contact relations and other features which since N.V. Ussing's comprehensive 1912 memoir, if at all treated, have been so in a cursory way. It is endeavoured to bring new information about composition and evolution of the Ilímaussaq agpaitic melts and to present a synthesis of the petrology of the complex.

Brief overview of the geology of the Ilímaussaq complex

The Ilímaussaq complex (Fig. 1) is one of the intrusive complexes of the Mid-Pro-

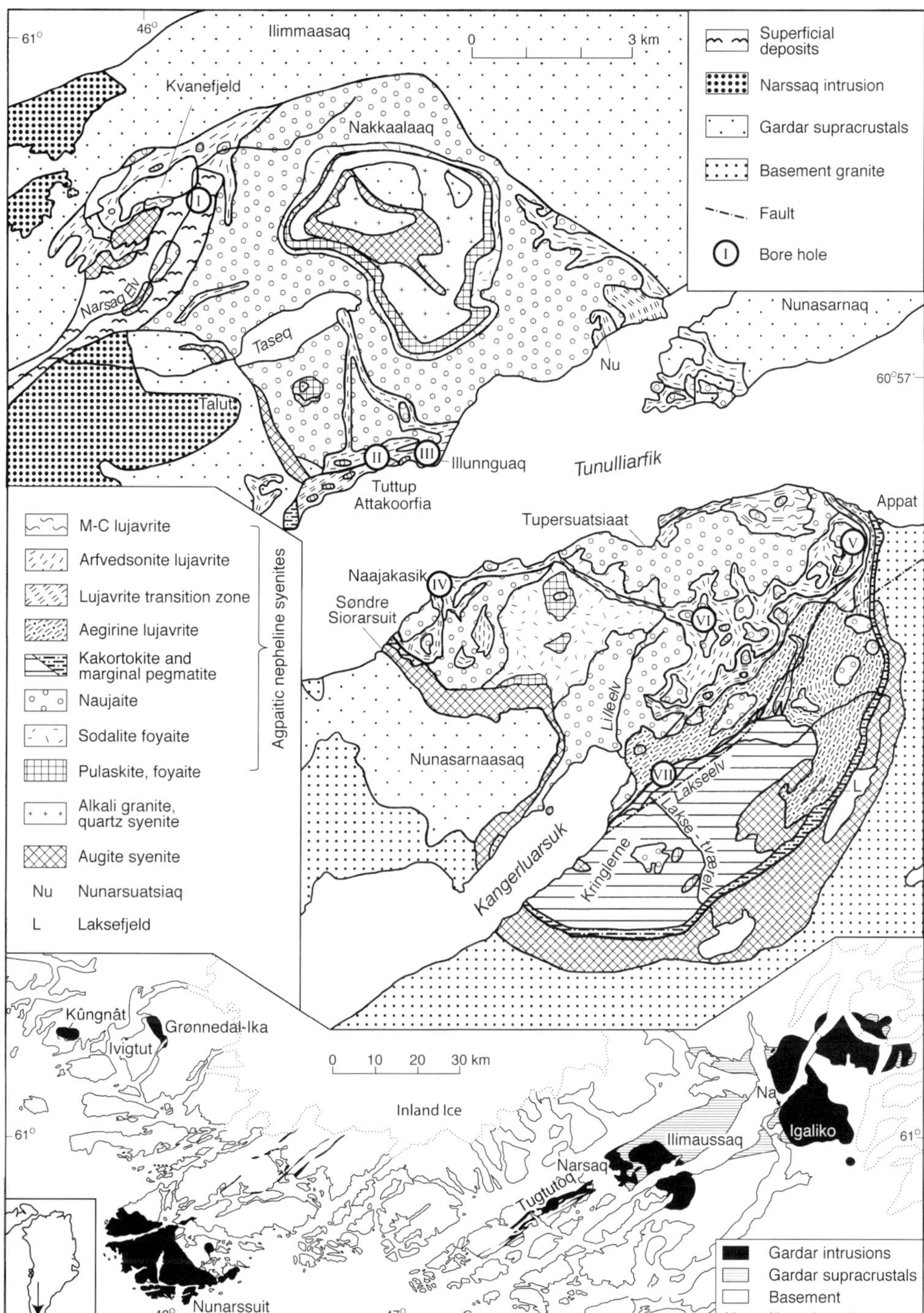

Fig. 1. Geological map of the Ilímaussaq complex based on Ferguson (1964) and Andersen et al. (1988) and sketch map of the Gardar igneous province.

terozoic Gardar igneous province in South Greenland (Upton *et al.* 2003). According to U-Pb and Rb-Sr dating the age of the complex is 1060 Ma ± 5 (Krumrei *et al.* 2006) or ±2.3 (Waight *et al.* 2002). It was formed during at least three intrusive stages:

(1) Augite syenite which is now only preserved as a partial rim around and in the roof of the complex and as xenoliths in rocks of stage 3.

(2) Alkali granite and alkali syenite which are preserved in the roof of the

Table 1. The major rock types of the Ilímaussaq complex

rock type	texture	essential minerals	minor minerals
augite syenite	hypidiomorphic to xenomorphic granular, massive or layered, medium to coarse	alkali feldspar, olivine, ferrosalite, ferropargasite	titanomagnetite, apatite, biotite, pyrrhotite, plagioclase (An ≤ 20)
pulaskite and foyaite	massive, medium to coarse, platy feldspars	alkali feldspar, fayalite, hedenbergite, aegirine augite, katophorite, nepheline	apatite, titanomagnetite, biotite, aenigmatite, fluorite, eudialyte
sodalite foyaite	foyaitic, coarse	alkali feldspar, nepheline, sodalite, aegirine augite, katophorite, fayalite	apatite, titanomagnetite, aenigmatite, eudialyte, rinkite, fluorite, biotite, steenstrupine
naujaite	poikilitic, coarse to pegmatitic	alkali feldspar, sodalite, nepheline, aegirine, arfvedsonite, eudialyte	rinkite, aenigmatite, fayalite, apatite, polylithionite, sphalerite, villiaumite, pectolite
kakortokite	laminated, layered, medium to coarse	alkali feldspar, nepheline, eudialyte, aegirine, arfvedsonite	sodalite, aenigmatite, rinkite, fluorite, löllingite
lujavrite	laminated, fine-grained, sometimes layered or massive, medium to coarse	microcline, albite, nepheline, sodalite, analcime, naujakasite, aegirine, arfvedsonite, eudialyte	steenstrupine, monazite, britholite, villiaumite, sphalerite, pectolite, lovozerite, vitusite, polylithionite, ussingite, neptunite
alkali granite, quartz syenite	hypidiomorphic granular, medium to coarse	alkali feldspar, quartz, aegirine, arfvedsonite	aenigmatite, elpidite, zircon, ilmenite, pyrochlore, neptunite, fluorite

complex and as xenoliths in the uppermost rocks of stage 3.

(3) Nepheline syenites which occupy the major part of the exposed volume of the complex and are divided into a roof series, an intermediate series and a floor series.

The roof series is made up of pulaskite, foyaite, sodalite foyaite and naujaite, the two last-named of agpaitic composition (see Table 1 for explanation of rock types). The intermediate series is sandwiched between the roof and floor series and consists of meso- to melanocratic agpaitic nepheline syenites described as lujavrites. The lower part of this series consists of aegirine-rich lujavrites, the upper part of arfvedsonite-rich lujavrites. The exposed part of the floor series is made up of in principle luja-

Fig. 2. Trough layering in kakortokite in the innermost part of the marginal pegmatite. The rock is characterised by thin eudialyte-rich layers. The trough is intersected by a fault. The dark rock on the upper left side of the trough is a microsyenite xenolith (see Fig. 12). The height of the exposure is about 2 m. South coast of Kangerluarsuk near the west contact of the complex.

vritic rocks. Ussing (1912), however, restricted the term lujavrite to the rocks of the present-day intermediate series and introduced the name kakortokite for the group of rocks which constitutes the floor series. Lujavrites and kakortokites are made up of the same major minerals: alkali feldspar, nepheline, arfvedsonite, aegirine and eudialyte, but have slightly different minor and accessory minerals. Kakortokites are layered medium- to coarse-grained, laminated or granular rocks; lujavrites are generally fine-grained and laminated. It is inferred that the unexposed part of the floor series was formed simultaneously with the major part of the roof series.

Fig. 3. Marginal pegmatite consisting of a homogeneous matrix and short thin pegmatite veins. North coast of Kangerluarsuk in the west contact of the complex.

The marginal pegmatite and other new sources of information about the magmatic evolution of the Ilímaussaq complex

Marginal pegmatite is the name given to a complicated mixture of a medium- to coarse-grained, massive-textured matrix of agpaitic nepheline syenites and intersecting thin and short pegmatitic veins. The matrix constitutes 40% to more than 60% of the marginal pegmatite and will be described below. The pegmatitic veins have a simple agpaitic mineralogy being composed of microcline, nepheline, sodalite, aegirine, arfvedsonite, aenigmatite and eudialyte and with biotite, rinkite, astrophyllite and fluorite as the most common accessory minerals. The marginal pegmatite forms a rim around the southern part of the complex from the west contact on the south coast of Kangerluarsuk to the east contact at Appat on the south coast of Tunulliarfik (Fig. 1). Along the west contact of the complex, there are short stretches of marginal pegmatite on the north coast of Kangerluarsuk and on the south and the north coasts of Tunulliarfik. A small occurrence is found in the east contact on the north coast of Tunulliarfik. The marginal pegmatite has, however, not been observed along most of the west, east and north contacts of the complex and it appears that it is restricted to the lower part of the complex made up of kakortokite and aegirine lujavrite (Fig. 1). As exceptions to this, marginal pegmatite is found near the roof of the complex at the Kvanefjeld plateau and at Nakkaalaak (Fig. 1). The different sections of the marginal pegmatite will be described separately below.

The marginal pegmatite along the south contact

The west contact of the complex on the south coast of Kangerluarsuk is the classic area for the examination of the marginal pegmatite (Ussing 1912; Ferguson 1964, 1970a; Bohse *et al.* 1971; Bohse & Andersen 1981). The marginal pegmatite forms an up to 100 m wide zone between the augite syenite rim and the layered kakortokite. The main mass of kakortokite consists of a repetition of units composed of black, red and white layers of cumulative agpaitic nepheline syenites. The layers pass laterally into the marginal pegmatite in such a way that layering persists but changes from the regular repetition of white, red and black layers in the main mass to thin layers which are folded and broken, are rich in eudialyte and form the matrix of the marginal pegmatite. At this place, the matrix shows cross-bedding, graded bedding and wash-out channels (Fig. 2), features which are absent from the main mass of kakortokite (Sørensen 1969; Bohse *et al.* 1971; Sørensen & Larsen 1987). The dip of the layers changes from almost horizontal in the regularly layered kakortokite to 40-50° inwards in the matrix of the marginal pegmatite close to the contact. Nearest to the kakortokite, the pegmatite veins are dyke-like and dominated by microcline and arfvedsonite. Towards the augite syenite they have irregular forms, more diffuse contacts and are rich in eudialyte. The thinly-layered matrix of the marginal pegmatite is the lowermost exposed part of the Ilímaussaq complex. It contains xenoliths of augite syenite rimmed by amphibole and clinopyroxene crystals. A large mass of slumped kakortokite has been mapped in the exposed layered kakortokite (Bohse *et al.* 1971; Andersen *et al.* 1988). A bore hole with an outward dip of 45° in the marginal pegmatite near the coast close to the west contact between marginal pegmatite and augite syenite did not reach the augite syenite but penetrated slumped kakortokite between 62 and 75 m in the core (Guttenberger & LeCouteur 1992).

At higher altitudes, the matrix rock of the marginal pegmatite encloses small and large augite syenite xenoliths and contains here and there masses of naujaite. One example is encountered at about 400 m above sea level (a.s.l.) in the south contact of the complex in the ridge north of the big lake in the augite syenite rim (Fig. 1). Two bore holes penetrated five metres of naujaite between marginal pegmatite and augite syenite (Guttenberger & LeCouteur 1992). In the naujaite here, sodalite crystals are poikilitically enclosed in large plates of microcline perthite and minor up to 5 cm long arfvedsonite crystals. In the west contact, a several metres wide naujaite mass occurs from ca. 100 to ca. 150 m a.s.l. It is separated from the augite syenite by pegmatite. The adjacent marginal pegmatite contains small naujaite masses which appear to be under disaggregation in the matrix. About 200 m a.s.l., the matrix rocks have features in common with kakortokite as well as with naujaite. They have crystals of microcline, eudialyte, arfvedsonite and aegirine, which poikilitically enclose crystals of sodalite and occasionally nepheline. The naujaite is here in contact with the augite syenite rim, only separated from this rock by a thin zone of pegmatite. Augite syenite xenoliths are separated from the matrix of the marginal pegmatite by naujaite (Steenfelt 1972). The naujaite appears to be older than the marginal pegmatite.

Fist-size masses of massive rocks with grain size 1-5 mm which occur between pegmatite minerals may be parts of the matrix of the marginal pegmatite or perhaps have been formed by quenching of the pegmatitic melts. One sample (104363A) will be described in the petrography section.

The marginal pegmatite is in contact with kakortokite from Kangerluarsuk to Laksefjeld, from there to near Appat with the overlying lujavrites (Fig. 1).

The marginal pegmatite between Kangerluarsuk and Tunulliarfik

On the north coast of Kangerluarsuk and the south coast of Tunulliarfik, marginal pegmatite occurs between augite syenite and aegirine lujavrite in the west contact of the complex (Andersen *et al.* 1988). Marginal pegmatite is lacking in the intervening area where naujaite and locally sodalite foyaite and pulaskite are in contact with augite syenite (Fig. 1). The contact between naujaite and augite syenite is only visible in a few places. It is generally covered by scree and crumbling augite syenite and naujaite. Where visible, augite syenite and naujaite have sharp contacts with no textural or compositional changes in either rock (Ferguson 1964).

On the north coast of Kangerluarsuk, the matrix of the marginal pegmatite is made up of a homogeneous, massive-textured and locally weakly laminated agpaitic nepheline syenite (sample 104361A). There is an up to two cm wide zone of a finer-grained rock (104361B) in contact with the augite syenite. The latter is strongly recrystallized near the contact. The matrix of the marginal pegmatite is intersected by short pegmatites (Fig. 3). The rock assemblage here, including the augite syenite, is intersected by thin lujavrite dykes and by

Fig. 4. Marginal pegmatite. Left hand side of the photo shows the homogeneous matrix with irregular network of pegmatite veins. Right hand side, the pegmatites (the dark bands) are arranged as parallel bands in the homogeneous matrix. West contact, north coast of Tunulliarfik.

zones of strong alteration containing albite, ilvaite, andradite, bavenite, fluorite, etc. (Petersen *et al.* 1995). Inside the marginal pegmatite there is a small outcrop of aegirine lujavite and after that follows the main mass of naujaite (Fig. 1).

On the south coast of Tunulliarfik, the marginal pegmatite in the west contact is identical to that on the north coast of Kangerluarsuk, but the massive-textured matrix and the pegmatitic veins are strongly altered by post-magmatic processes. The primary minerals have been replaced by epidote, hematite, grossularite-andradite, albite, fluorite and ilvaite (Ussing 1912). At this place, the marginal pegmatite only occurs between augite syenite and aegirine lujavrite in the coast exposure. Marginal pegmatite and aegirine lujavrite both terminate upward where naujaite is in direct contact with augite syenite (Andersen *et al.* 1988).

The marginal pegmatite at the north coast of Tunulliarfik

A zone of marginal pegmatite occurs between aegirine lujavrite and the basaltic country rocks in the west contact of the complex (Figs. 1, 4, 5). It is marked as an agpaitic dyke on the map of Ferguson (1964). The basalt is strongly contact metasomatised and is intersected by nepheline syenite dykes. There is a thin pegmatite vein along the contact between basalt and marginal pegmatite.

The matrix of the marginal pegmatite is a homogeneous, medium- to coarse-grained, massive-textured rock (109303) locally showing weak lamination. It is intersected by randomly oriented thin and short pegmatitic dykes (Fig. 4), but about 100 m from the contact the pegmatites form parallel bands, up to 0.5 m thick (Fig. 5), i.e. a distribution similar to the pegmatite veins in the marginal pegmatite on the south coast of Kangerluarsuk. The almost vertical contact between the marginal pegmatite and the intruding aegirine lujavrite is sharp and intersects the pegmatite bands. The lamination of the aegirine lujavrite is parallel to the contact on the marginal pegmatite but away from the contact the aegirine lujavrite is strongly deformed and contains xenoliths of basalt, augite syenite and naujaite. Thus, the lujavrite

Fig. 5. The same place as Fig. 4 but slightly more to the east (to the right). The marginal pegmatite with parallel pegmatite bands is intersected by aegirine lujavrite (dark green). The distinct lamination of the lujavrite is parallel to the nearly vertical contact. The background mounatain wall is seen en face in Fig. 23.

has penetrated and disrupted the marginal contact of the earlier augite syenite intrusion.

Marginal pegmatite has also been discovered in the east contact of the complex at Nunasarnaq (Fig. 1), where it wedges out upwards between aegirine lujavrite and the strongly sheared country rocks which here consist of Gardar sandstone overlain by basalt (Fig. 6). On the inner side of the marginal pegmatite follows aegirine lujavrite, a sequence of alternating thin layers of aegirine lujavrite and arfvedsonite lujavrite, a thicker layer of aegirine lujavrite and then layered arfved-

Fig. 6. The east contact of the complex at Nunasarnaq. On the right side: Gardar sandstone (SS) showing vertical shearing. The marginal pegmatite (MP) wedges out upwards and is followed inwards by aegirine lujavrite (GL, brownish).

Fig. 7. Quenched pegmatite (107728). A fine-grained eudialyte-rich rock with cavities occupies the space between pegmatitic microcline (white) and arfvedsonite crystals (black) The specimen is 12 cm long. Head of Kangerluarsuk at the outlet of Lilleelv.

sonite lujavrite with a weak westerly dip.

The marginal pegmatite at Kvanefjeld and Nakkaalaak

At Kvanefjeld in the northernmost part of the complex (Fig. 1), lujavrites and naujaite are in contact with the volcanic rocks of the roof and to the north of the complex. Along a part of the north contact, the contact facies of the naujaite is rich in pegmatite veins which recall the marginal pegmatite at lower levels in the complex (Sørensen *et al.* 1969; Steenfelt 1972, 1981). Locally the naujaitic matrix of this marginal pegmatite contains areas of homogeneous, massive-textured rocks which have black, red and white patches enriched in respectively arfvedsonite, eudialyte and microcline but without forming layers. There appears to be a gradual transition from naujaite to the massive-textured rocks and *vice versa*. In the massive-textured rock, small crystals of sodalite and occasionally nepheline are poikilitically enclosed in larger crystals of eudialyte, aegirine, arfvedsonite and aenigmatite which are set in fine-to medium-grained rock made up of microcline, nepheline, aegirine and eudialyte (104380A). The pegmatites form lenticular bodies or an irregular network of short veins and show a gradual transition to naujaite, but sharp contacts against the massive-textured rock.

At Nakkaalaak, naujaite in contact with the basaltic roof has developed a zone of marginal pegmatite which sends apophyses into the basalt.

Quenched pegmatitic melts

Pegmatites in especially naujaite and the marginal pegmatite contain here and there masses of fine-grained, massive-textured rocks measuring up to ten cm or more which occupy interstices between pegmatite minerals. Some of the fine-grained masses are rich in cavities. The immediate interpretation of the mode of occurrence, the fine grain-size and the texture of these masses is that they have been formed by rapid cooling of pegmatitic melts, possibly initiated by a release of volatiles, and that their chemical composition may be equivalent to the composition of their parent melts.

Sample 107728 (Fig. 7) was collected in a near horizontal pegmatite sheet in naujaite located at the head of Kangerluarsuk immediately to the north of the outlet of Lilleelv (Fig. 1) which was described by Ussing (1912) and Sørensen (1962). The pegmatite is asymmetrically zoned with a lower zone rich in eudialyte and an upper zone of large crystals of microcline, aegirine and arfvedsonite. Sample 107728 was collected in the upper zone and consists of large crystals of microcline, aegirine and arfvedsonite. The fine-grained rock, which is full of holes, occurs as interstitial masses measuring up to four centimetres and branching out as thin films along the grain boundaries of the pegmatite. It is thought to have been formed by rapid consolidation of a residual melt which occupied open spaces between the pegmatite minerals.

Sample 104363A was collected in the west contact of the complex south of

Kangerluarsuk at about 200 m a.s.l. It is a fist-size mass, about 5 cm in diameter, of a massive-textured rock which occurs in a pegmatitic segregation (Fig. 8). It may be a part of the matrix of the marginal pegmatite or represent a quenched pegmatitic melt.

A mini-scale model of the agpaitic magma chamber of the Ilímaussaq complex

Sodalite is an important constituent of the roof series of the Ilímaussaq complex. It is intergranular in sodalite foyaite and the liquidus phase in naujaite which since Ussing (1912) has been interpreted as a flotation cumulate of sodalite. It is therefore of some interest to note that pegmatites in naujaite and kakortokite show an enrichment in sodalite in their upper parts and mimic the general structure of the Ilímaussaq complex.

Some lenticular pegmatite sheets in naujaite show an asymmetrical zonation (Ussing, 1912; Sørensen 1962, Sørensen & Larsen 1987). Their upper part consists of sodalite crystals which measure one centimetre or more across. This zone is underlain by a blocky zone dominated by large microcline and arfvedsonite crystals. The lower part is rich in eudialyte and passes gradually into the underlying naujaite. The contact to the overlying naujaite is fairly sharp. When compared with the Ilímaussaq complex, the sodalite-rich upper part corresponds to the naujaite of the roof series, the lower part to the kakortokites. Between the microcline and eudialyte zones there are masses of aegirine-rich rocks which with regard to mineralogical composition recall the lujavrite sandwich horizon of the complex and like the lujavrites may be rich in rare elements (Sørensen 1962).

The layered units of the main mass of kakortokite have been numbered: units 0 to +17 form the upper part, units -1 to -11 the lower part of the exposed kakortokite layered sequence (Bohse *et al.* 1971). The bottom of the kakortokites is not exposed. The units consist of a lower up to 1 m thick black layer rich in arfvedsonite, an intermediate 0 to 1 m thick red layer rich in eudialyte and an upper about 8 m thick white layer rich in white alkali feldspar. The boundaries between units are sharp; the boundaries within units are gradual.

On the south coast of Kangerluarsuk, about 100 m from the marginal pegmatite, a roughly 10 cm thick pegmatite sheet occurs between kakortokite layer -10 black and the underlying layer -11 white. The upper surface of the sheet is wavy; the lower surface is plane and parallel to the contact between the two layers. The pegmatite passes into the underlying kakortokite over a few centimetres. The pegmatite sheet is rich in sodalite in its upper part and rich in arfvedsonite in its lower part. In one place, the pegmatite protrudes upward and forms a cupola which measures 25 cm from base to top (Fig. 9). Its lateral contacts against the black kakortokite intersect the horizontal lamination of the kakortokite. Arfvedsonite crystals grow on the marginal contact of the cupola

Fig. 8. Fine-grained, massive and compact rock (104363A) which contains about 40 vol.% aegirine and arfvedsonite (see also Figs. 17a, b). It is in contact with a eudialyte-rich pegmatite with some nepheline and interstitial arfvedsonite. West contact, south coast of Kangerluarsuk.

Fig. 9. The cupola of the mini-scale model pegmatite protruding into the lower part of kakortokite black layer -10. Note the black rim between pegmatite and kakortokite, the zonation of the pegmatite, light coloured in upper part (see also Figs. 10, 18a, b), darker coloured in lower part (see also Figs. 11, 19) and with small fine-grained black bodies in the lower part (see also Fig. 11). South coast of Kangerluarsuk.

Fig. 10. Upper part of cupola of mini-scale model pegmatite (107724) showing the upper sodalite-rich part (see also Figs. 18a, b) and increasing content of eudialyte downward in the rock. Same place as Fig. 9.

and protrude into the kakortokite parallel to its lamination. Sodalite and cm-large arfvedsonite crystals have been introduced into the kakortokite immediately above the pegmatite. The cupola shows mineral segregation, an upper part rich in sodalite, a middle part of sodalite and eudialyte (Fig. 10) and a lower laminated part with microcline, arfvedsonite and eudialyte (Fig. 11).

Lenticular bodies of a black fine-grained to aphanitic rock occur in the lower part of the sodalite-eudialyte zone and in the upper part of the microcline-arfvedsonite-eudialyte zone. They have sharp contacts against the host rock, are up to three cm thick and taper out laterally more or less parallel to the lamination of the host rock, reaching lengths of up to eight cm (Fig. 11). One sample (107724) has been analysed.

If this pegmatite cupola is considered a mini-scale model of the Ilímaussaq complex, the sodalite-rich upper part corresponds to the roof series, the laminated lower part to the floor series. The aphanitic black rock could then correspond to the lujavritic intermediate series.

Xenolith in kakortokite

The kakortokites contain xenoliths of augite syenite, naujaite and rare foyaite which are mainly found in layer +3, probably the result of a major collapse of wall and roof rocks (Bohse *et al.* 1971). Xenoliths of extraneous origin have not been recorded until now. One such xenolith has, however, been found in the transition zone between the marginal pegmatite and the regularly layered kakortokite a few metres from the south coast of Kangerluarsuk. The transition zone is characterised by very thin layers, cross-bedding, trough-layering and deformation of layers. The xenolith is box-shaped and the exposed surface area is 50 x 40 cm. It occurs on the edge of the most prominent trough structure of the area (Fig. 2). Its horizontal sides are parallel to the kakortokite layering; the vertical sides are at right angle to the layering. The xenolith consists of microsyenite with scattered large aegirine crystals. It is separated from the kakortokite by a pegmatitic zone rich in prismatic aegirine crystals (Fig. 12).

Petrography

The matrix of the marginal pegmatite

Samples of homogeneous-looking matrix rocks of the marginal pegmatite from the south and north coasts of Kangerluarsuk, the north coast of Tunulliarfik and from Kvanefjeld were selected for closer study and chemical analysis. These samples have massive textures characterised by randomly arranged plates of microcline. The grain size varies from 0.2 to 0.5 cm, the feldspar plates may be up to one cm long.

Sample 104361A from the north coast of Kangerluarsuk (Figs. 13, 14a, b) is dominated by a framework of translucent plates of microcline which are rimmed and partially intergrown by albite. In hand specimen, nepheline crystals are easily recognised and are estimated to make up about 15 vol. % of the rock. In thin section, unaltered nepheline grains are extremely rare, the major part of the grains are pseudomorphed by analcime crowded with tiny flakes of sericite. Green clinopyroxene occurs as tiny interstitial prismatic grains and as up to one cm long crystals which have pale green central parts crowded with pigmentation and more vividly green marginal parts. They have blebs of fluorite. Amphibole crystals of deep green colour are almost completely replaced and rimmed by brown aegirine and by brownish-black pigmentary material. The larger pyroxene and amphibole grains poikilitically enclose microcline plates, altered eudialyte and analcime crystals which most probably are altered nepheline grains. Eudialyte pseudomorphs consist of aggregates of minute zircon crystals embedded in brownish-black pigmentary material, natrolite, aegirine, analcime and fluorite. This is the type of alteration of eudialyte which was described by Ussing (1894). Fluorite also occurs as independent grains and as blebs in the altered amphibole. Analcime is an interstitial mineral.

Fig. 11. Lower part of cupola of the mini-scale model pegmatite. Left side of photo shows the laminated rock in the lower part of the cupola and small body of the fine-grained black rock, see also Fig. 19. Right side shows the overlying eudialyte-sodalite-rich rock (see Fig. 10). Same place as Fig. 10.

Fig. 12. Fine-grained microsyenite xenolith in kakortokite. Note the pegmatitic rim and the aegirine crystals. Same place as Fig. 2.

Fig. 13. The matrix of the marginal pegmatite (104361A) from the west contact on the north coast of Kangerluarsuk, see also Fig. 3. The pinkish spots are altered nepheline grains.

The marginal pegmatite has developed an about two centimetres thick fine-grained zone (104361B) in contact with augite syenite, one of the few examples of a chilled zone in the Ilímaussaq agpaitic rocks. The grain size is 1-2 mm whereas it is 2-5 mm away from the contact. The contact rock is made up of the same minerals as away from the contact. Translucent plates of microcline are arranged at an acute angle to the contact and are partially replaced by albite. Prismatic amphibole crystals are perpendicular to the contact. Nepheline is generally replaced by analcime and sericite. The pseudomorphs after nepheline form clusters. Small crystals of a pale green clinopyroxene are filled with black pigmentation and are interstitial to the microcline. Unaltered grains of a bluish-green amphibole are rare, most grains of this mineral are substituted by black pigmentation and are often rimmed by aegirine. Pseudomorphs after eudialyte crystals are similar to those described from 104361A. Albite and analcime are interstitial minerals. The adjacent augite syenite is strongly recrystallized.

Sample 109303 from the north coast of Tunulliarfik contains 10 to more than 50 vol. % nepheline and about 20% in the analysed material. The nepheline encloses small laths of feldspar and is partially altered to analcime and sericite. The first step in the alteration is the formation of brown pigmentation in the marginal parts of the grains. Unaltered plates of microcline, which are partially intergrown with albite, make up about 60 vol. %. There are small prismatic grains and larger grains of green clinopyroxene. Some of the larger pyroxene

Fig.14 a, b Microphotos of whole thin section of 104361A, a. plane polarised light, b. crossed polarised light. Note the texture dominated by the microcline tablets and the dark interstitial grains which are altered nepheline, arfvedsonite and eudialyte and secondary aegirine. The long side of the thin section measures 3.5 cm.

grains have disintegrated into aggregates of smaller grains of pale green clinopyroxene. Amphibole is completely altered to dark pigmentary material. The amphibole pseudomorphs have partial rims of pale green clinopyroxene. The large pyroxene grains and the altered amphibole grains poikilitically enclose well-preserved nepheline and analcime crystals. The latter may represent altered nepheline or perhaps sodalite. Eudialyte pseudomorphs are similar to those described from 104361A, B but additionally contain a carbonate mineral. Analcime is an interstitial mineral.

Sample 104380A from Kvanefjeld (Fig. 15) contains 30-40 vol. % nepheline crystals in a matrix of microcline and minor eudialyte, arfvedsonite and green aegirine. The arfvedsonite grains grow between the other minerals, a first step in the formation of poikilitic intergrowth (Fig. 16a, b). Arfvedsonite, aegirine and aggregates of eudialyte crystals poikilitically enclose crystals of nepheline and small aegirine crystals. Some aegirine crystals have cores of arfvedsonite and some arfvedsonite crystals have cores of aenigmatite. Nepheline is partially altered to natrolite. Microcline is rimmed by albite. Arfvedsonite and green aegirine are partially altered to brown aegirine. Eudialyte occurs as unaltered crystals or is partially altered to catapleiite. There are interstitial patches of natrolite, white mica and pectolite.

Fig. 15. Hand specimen of 104380A: the matrix of the marginal pegmatite at Kvanefjeld, see also Figs. 16a, b.

Fig. 16a, b. Microphoto of whole thin section of 104380A. Note the texture and the interstitial nature of the arfvedsonite grains (black) and to a minor extent the aegirine grains (green). The felsic minerals, among which abundant nepheline crystals, are very clear, i.e. very little altered. a. plane polarised light, b. crossed polarised light. The long side of the thin section measures 3.7 cm.

Fig. 17a, b. Microphoto of whole thin section of 104363A, fine-grained rock enclosed in pegmatite, see also Fig. 8. The large pegmatite grains are nepheline and eudialyte. Note the texture of the fine-grained rock dominated by microcline tablets, the fairly high content of mafic minerals, the long prismatic grains of aegirine (green in a) and the unaltered condition of all minerals. a. plane polarised light, b. crossed polarised light. The long side of the thin section measures 3.8 cm.

Sample 104363A is a massive-textured rather compact rock associated with a pegmatitic segregation of eudialyte and nepheline crystals from the west contact of the complex south of Kangerluarsuk (Figs. 8, 17a, b). Its grain size is 1-5 mm. It contains about 40 vol. % mafic minerals and about 10% eudialyte. The texture is dominated by randomly arranged plates of microcline. Corroded nepheline crystals are larger than the feldspar plates. The eudialyte crystals are less than 1 mm across. Slender prismatic aegirine crystals have swallow tails and enclose small feldspar grains along their median line. Irregularly-shaped arfvedsonite grains rimmed by aegirine grow between and enclose the other minerals in a fashion which suggests an early stage in the development of the poikilitic texture of naujaite. Analcime forms thin rims on nepheline but is otherwise practically absent. The fine grain size, the texture of the microcline plates and the inclusions and swallow tails of aegirine are indications of rapid cooling of this rock.

Quenched pegmatites

The above-mentioned sample 104363A may be an example of a quenched pegmatitic melt. An additional example is sample 107728 (Fig. 7) in which a fine-grained rock is interstitial to pegmatite minerals. It consists of 40-60% microcline, 20-30% aegirine and 20-30 vol. % eudialyte (Fig. 7). The grain-size is 1-2 mm. The texture is dominated by randomly arranged microcline plates with interstitial aegirine and eudialyte crystals. The last-named crystals form clusters which are located in the marginal parts, whereas very little eudialyte is found in the central parts of the fine-grained rock. Small corroded nepheline grains are rimmed by analcime. Minor components are albite, neptunite, pectolite and white mica. A third example will be described in the next section.

Fig. 18a, b. Microphotos of whole thin section of 107772B, the sodalite-eudialyte-rich upper part of the mini-scale model pegmatite, see also Figs. 9, 10. Note the granular texture, the very low contents of mafic minerals and feldspar. Eudialyte crystals have faint red colour in 18a and are greyish in 18b. a. plane polarised light, b. crossed polarised light. The long side of the thin section measures 3.6 cm.

The mini-scale model pegmatite

The cupola of this pegmatite shows a distinct mineral segregation (Figs. 9-11, 18a, b, 19a, b). Its uppermost part is greenish-grey and consists almost entirely of sodalite crystals which measure 0.1 to 0.5 cm in diameter. This sodalitolite zone is up to 10 cm thick in the apical part of the cupola. Here and there corroded nepheline crystals are enclosed in the sodalite. There are interstitial aggregates of small albite laths and in its lower part, subordinate mm-size eudialyte crystals and very scarce interstitial arfvedsonite and aegirine which have inclusions of albite.

A rather sharp but wavy contact and a thin albite zone separate the sodalite zone from the underlying speckled rock which is composed of sodalite and mm-size eudialyte crystals, the amount of the last-named mineral increasing downward (Figs. 10, 18a,b). The sodalite crystals are interstitial to the eudialyte crystals and there are interstitial aggregates of albite laths. The top sodalite zone and the underlying sodalite-eudialyte zone both bend around and taper out downward along the lateral contacts of the cupola.

The sodalite-eudialyte zone is with rather sharp contact underlain by a rock composed of microcline plates and eudialyte crystals with abundant interstitial arfvedsonite which together display horizontal lamination (Figs. 19a,b). Small aegirine crystals are enclosed in large amphibole grains. Aegirine also occurs as large prismatic grains. Additional minor minerals are nepheline, albite, analcime and fluorite, the last-named mineral as interstitial crystals and as blebs in amphibole. There are clusters of sodalite crystals akin to the sodalite zone in the apical part of the cupola. The sodalite encloses corroded nepheline grains.

The black aphanitic to fine-grained rock in the lower part of the cupola (Fig. 11) consists of microcline laths and inter-

Fig. 19. Whole thin section of the lower, laminated part of the mini-scale model pegmatite, 107723. The rock is dominated by microcline plates, arfvedsonite prisms and eudialyte crystals (pinkish in a). a. plane polarised light, b. crossed polarised light. See also Fig. 11. The long side measures 3.4 cm.

stitial arfvedsonite. Additional minerals are nepheline, aegirine, aenigmatite and interstitial albite and analcime. Eudialyte has not been identified. The lamination of the rock bends around nepheline crystals and rounded analcime grains. The rock is rich in fluorite as independent crystals and as blebs in amphibole. The form of the black bodies and their relationship to the minerals of the surrounding pegmatite exclude that they are xenoliths. They are considered to have been formed by rapid cooling of residual pegmatitic melts trapped in the central and lower part of the cupola. A chemical analysis of the black rock (107724) is presented in Table 2.

The distribution of the components of the cupola recalls the overall distribution of the agpaitic rocks of the Ilímaussaq complex: the upper sodalite-rich part corresponds to the naujaite, the bottom laminated part to the kakortokite and the black bodies to the lujavrites. The cupola may be considered a mini-scale version of the agpaitic part of the Ilímaussaq complex.

Xenolith in kakortokite

The xenolith is made up of a fine-grained greyish-black rock with white spots and up to 10 cm long aegirine crystals.

In thin section, the rock is dominated by randomly arranged acicular aegirine in a matrix of faintly turbid microcline which displays the penetration type of twinning characteristic for agpaitic rocks. Irregularly shaped, lucid albite grains dominate parts of the rock. There are tiny crystals of zircon, a brown isotropic mineral, fluorite and around cavities small black grains. The rock contains sector-zoned aegirine crystals which may be several centimetres long.

The rock resembles the microsyenite dykes and sheets which intrude kakortokite, aegirine lujavrite and naujaite (Rose-Hansen & Sørensen 2001). A microsyenite sheet in naujaite adjacent to bore hole I in the bed of Narsaq Elv (Fig. 1) shows the same type of albite replacing microcline as is seen in the xenolith.

Sample 104352 of the microcline-rich variety has been analysed (Table 2).

Whole-rock chemistry

A major geochemical study of the Ilímaussaq complex directed by J.C. Bailey is in progress. It is based on 120 large samples of the major rock types of Ilímaussaq. The rocks and their minerals are being analysed for about 60 elements. A preliminary account of the project was presented by Bailey *et al.* (2001).

Table 2. Chemical analyses of homogeneous matrix rocks from the marginal pegmatite (M.P.), of dykes intersecting the country rocks, of a xenolith in kakortokite, and a microsyenite sheet. Major elements in weight percent, trace elements as ppm if not otherwise indicated.

	104361B matrix M.P.	**104361A matrix M.P.**	**109303 matrix M.P.**	**104380A matrix M.P.**	**104363A matrix M.P.**	**153099 dyke**	**153200 dyke**	**107724 Black body in pegmatite**	**104352 Xenolith in kakortokite**	**109330 Micro-syenite**
SiO_2	54.10	54.60	52.34	49.50	53.69	56.71	52.98	53.90	60.69	62.39
TiO_2	0.20	0.20	0.10	0.28	0.30	0.86	0.47	0.50	0.34	0.51
ZrO_2	1.02	0.75	0.98	1.89	0.84	0.14	0.35	0.06	0.46	0.10
Al_2O_3	15.72	14.53	19.05	19.01	13.19	15.81	16.55	16.65	13.01	15.03
Fe_2O_3	7.96	8.50	6.95	5.29	8.06	2.42	5.00	3.85	7.86	5.30
FeO	2.43	2.40	0.22	1.54	4.03	7.12	4.27	6.21	0.67	0.41
MnO	0.26	0.24	0.26	0.25	0.46	0.31	0.44	0.23	0.13	0.11
MgO	0.26	0.16	0.12	0.13	0.23	0.42	0.35	0.21	0.13	0.10
CaO	2.64	2.78	1.08	1.83	1.32	2.46	1.89	2.22	0.40	1.61
Na_2O	6.79	7.16	9.89	13.27	10.70	7.15	9.65	8.58	5.21	6.99
K_2O	4.09	5.27	4.56	2.88	3.08	5.25	5.24	5.79	9.30	6.67
Cl	0.03	0.03	0.03	0.27	0.14	0.09	0.46	0.29	0.01	0.01
F	0.23	0.88	0.34	0.66	0.14	0.19	0.70	0.72	0.66	0.80
P_2O_5	0.05	0.05	0.09	0.07	0.10	0.42	0.23	0.01	0.04	0.09
vol/H_2O	2.93	2.51	3.56	2.45	2.51	0.60	1.42	1.06	0.34	0.47
	98.71	100.06	99.57	99.28	98.79	99.95	100.00	99.20	99.25	99.68
- O	0.11	0.38	0.15	0.34	0.09	0.10	0.40	0.37	0.28	0.34
total	98.60	99.68	99.42	98.94	98.70	99.85	99.60	98.83	98.97	99.34
A.I.	1.0	1.2	1.1	1.3	1.6	1.1	1.3	1.2	1.4	1.3
Cs	3.8	10.0	22.2	9.8	16.1	2.3	6.35	4.5	0.4	0.3
Rb	456	517	595	514	333	164	308	714	1260	363
Ba	379	365	111	78	185	137	136	237	266	219
Pb	62	56	117	61	108	27	61	25	183	9
Sr	269	166	396	109	81	36	111	37	22	32
La	541	403	741	855	488	147	341	26	474	158
Ce	930	770	1220	1550	785	312	628	49	580	313
Nd	426	295	493	590	279	129	202	23	154	157
Sm	79	58.7	81.7	99.5	51	21.5	38.0	3.7	15.1	31
Eu	7.3	5.8	8.1	10.7	4.8	2.1	3.0	0.3	2.1	2.9
Tb	13.6	10.0	15.8	21.2	9.1	2.6	4.4	0.5	3.1	3.9
Yb	50.2	36.3	47.0	70.2	38.1	9.1	14.9	2.3	19.6	8.5
Lu	7.0	5.0	6.5	9.9	5.4	–	–	0.4	3.0	1.2
Y	471	322	437	717	361	105	173	21	212	95
U	14	12	13	29	20	4.7	25	1.3	122	2.2
Th	50	31	31	54	60	17	59	4.7	210	7.4
Zr	8145	5549	7250	1.40%	6275	1014	2623	488	4170	924
Hf	184	128	186	331	127	20	48	11	54.5	–
Nb	830	528	746	1440	628	219	737	71	1110	28
Ta	54.7	34.0	49.8	127	30.7	–	–	3.5	21	–
Zn	300	227	395	491	1052	–	–	344	548	81
Sc	7	1.4	0.7	0.9	7	–	–	6	0.2	5.4
Ga	87	71	99	88	84	–	–	66	136	44

A.I. = agpaitic index – = not analysed

104361B and A, 109303, 104380A, 104363A: homogeneous matrix rocks; 153099 and 153200: dykes intersecting country rocks (from Larsen 1979); 107724: black body in mini-scale pegmatite; 104352: xenolith in kakortokite; 109330: microsyenite sheet (from Rose-Hansen & Sørensen 2001).

Table 3. CIPW weight norms and special norms of massive-textured rocks from the marginal pegmatite, the Ilímaussaq complex.

	104361A	109303	104380A	104363A
ap	0.11	0.15	0.15	0.25
il	0.39	0.14	0.61	0.56
z	1.20	1.52	2.81	1.29
or	31.62	28.16	17.53	18.25
ab	33.67	27.63	16.73	25.77
ne	7.73	26.18	36.74	14.39
ac	13.68	10.12	15.74	24.13
ns	–	–	2.67	3.05
wo	0.27	1.03	1.65	–
en	0.40	0.32	0.31	0.59
fs	4.59	–	2.87	8.05
hm	3.87	2.74	–	–
mt	–	1.48	–	–
hl	–	–	0.48	0.26
fl	2.49	0.73	1.40	0.29
Total	99.97	100.20	99.73	99.78
P_2O_5	0.05	0.06	0.06	0.10
MgO	0.16	0.13	0.12	0.24
TiO_2	–	–	0.32	–
F	–	0.16	0.45	0.14
Cl	–	–	0.29	–
z	–	0.34	–	–
eud	9.39	7.54	15.84	11.24
or	31.88	28.06	17.44	18.67
ab	36.96	30.82	13.50	24.62
ne	4.97	23.99	38.29	15.06
ac	–	1.60	9.90	14.76
arf	4.30	–	–	12.71
ns	–	–	–	–
wo	1.03	–	0.35	0.33
fs	–	–	1.35	–
hm	6.67	6.64	2.06	2.09
mt	2.31	–	–	–
fl	1.83	0.39	–	–
total	99.02	99.73	99.73	99.96

A general discussion of the geochemical evolution of the complex must await publication of the vast amount of data produced in this project. The present discussion will be restricted to data which elucidate the position of the homogeneous matrix rocks of the marginal pegmatite and the quenched pegmatites.

Major elements were analysed at the chemical laboratories of the Geological Survey of Denmark and Greenland (GEUS), Copenhagen, by XRF analysis of glass discs prepared with a sodium tetraborate flux (Kystol & Larsen 1999). Exceptions to this are: Na analysed by atomic absorption, FeO by a modified vanadate technique, and loss on ignition by gravimetry. Trace elements were analysed directly on pressed powder pellets by XRF using the techniques of Norrish & Chappel (1977) and F was analysed by the specific-ion electrode at the Geological Institute, University of Copenhagen. REE, Cs, Hf, U, Ta and Sc were analysed by INAA by Tracechem A/S, Copenhagen and by ICP-MS by GEUS.

Six homogeneous-looking matrix rocks and quenched pegmatites were analysed: 107724 and 104363A from the south coast and 104361A and B from the north coast of Kangerluarsuk, 109303 from the north coast of Tunulliarfik, and 104380A from Kvanefjeld (Tables 2, 5). Norms of four of these rocks are presented in Table 3. The average composition of the Ilímaussaq nepheline syenites and analyses of rocks that may represent liquid compositions are presented in Table 4, analyses of the major rock types of the agpaitic part of the Ilímaussaq complex in Table 5, and selected element ratios in Table 6.

It should be pointed out that most of the Ilímaussaq agpaitic rocks are cumulates. However, sodalite foyaite and parts of the foyaite have foyaitic texture and appear to have been formed by *in situ* crystallization of melts. These features must be kept in mind in the discussion of the whole-rock analyses of Tables 4, 5 and 6. There is a considerable variation in chemistry within each rock type but only one analysis of each type is considered in Tables 5, 6. For Table 5, unpublished analyses and analyses quoted from Bailey *et al.* (2001) have been chosen instead of the average com-

Table 4. Proposed average compositions of the agpaitic nepheline syenites of the Ilímaussaq alkaline complex together with analyses of sodalite foyaite from the complex, a phonolitic dyke in its immediate neighbourhood and interstitial glass in a nepheline syenite xenolith from Tenerife (Figures in brackets indicate number of samples).

	Average composition	**Sodalite foyaite**	**Average composition (29)**	**Sodalite foyaite (2)**	**Average composition (67)**	**Cl-poor sodalite foyaite (3)**	**Fe-rich phonolite**	**Interstitial glass, Tenerife**
SiO_2	51.86	49.38	50.66	49.97	49.79	51.01	51.83	49.44-55.22
TiO_2	0.33	0.63	0.35	0.54	0.33	0.34	0.55	1.27-1.52
ZrO_2	0.79	0.62	0.75	0.43	0.72	0.33	0.57	1.56-1.73
Al_2O_3	18.35	17.31	17.57	18.35	18.72	17.38	14.57	13.20-14.33
Fe_2O_3	6.50	4.20	5.86	4.96	5.02	4.73	7.56	4.90-5.86[1)]
FeO	3.68	5.25	3.77	4.04	3.58	4.62	4.61	
MnO	0.23	0.08	0.25	0.15	0.23	0.25	0.48	0.90-1.12
MgO	0.08	0.53	0.38	0.43	0.22	0.13	0.14	0.30-0.45
CaO	1.13	2.23	1.09	1.79	1.40	1.97	2.54	0.58-0.73
Na_2O	13.24	13.87	12.65	12.67	12.99	10.08	8.81	13.90-16.37
K_2O	2.91	2.55	3.40	3.17	3.67	3.93	4.87	3.39-3.93
Cl	1.13	1.68	1.11	1.79	1.40	0.09	0.33	0.23-0.36
F	–	–	0.19	0.29	0.25	0.41	0.84	0.89-1.35
P_2O_5	–	–	0.14	–	0.07	0.05	0.08	0.12-0.20
vol/H_2O	–	1.46	2.27	1.70	1.97	4.35	2.03	–
	100.25	99.78	100.48	99.89	100.34	99.98	99.81	92.90-100.91
-O		0.38	0.34	0.44	0.41	0.20	0.86	0.44-0.63
		99.40	100.14	99.89	99.93	99.78	98.95	92.46-100.28
agpaitic index	1.4	1.5	1.4	1.3	1.4	1.2	1.4	1.9-2.1
	Ussing (1912)[2)]	Ussing (1912)	Gerasimovsky (1969)[3)]	Gerasimovsky (1969)	Ferguson (1970a)	Bailey *et al.* (2001)	Larsen & Steenfelt (1974)	Wolff & Toney (1993)

[1)] Total iron

[2)] Calculated on a water-free basis under the assumption of the following proportions of rocks: Naujaite 30%, lujavrite 30%, kakortokite 10%, sodalite foyaite 10%

[3)] based on the same proportions as Ussing's calculation

positions of the individual rocks published by Ussing (1912), Gerasimovsky (1969) and Ferguson (1970a, b). The reason for this choice is that they show both major and trace elements and have been made by the same methods in the same laboratories as the new analyses.

The fact that most Ilímaussaq rocks are cumulates and that they commonly are extremely coarse-grained makes sampling of rocks that may represent melt compositions very difficult. The massive-textured rocks examined in this paper appear to have been formed by *in situ* crystallization of melts and may thus provide some information about magma compositions. These rocks have, however, been altered to various degrees by late-magmatic processes. This applies to most other loosely packed Ilímaussaq nepheline syenites, such as the white kakortokite, foyaite and sodalite foyaite, which have retained a major part of the interstitial residual melts, a

Table 5. Chemical analyses of homogeneous matrix rocks from the marginal pegmatite and of the major rock types from the Ilímaussaq complex. Major elements in weight percent, trace elements as ppm if not otherwise indicated.

	104361A	109303	104380A	154314	154346	154350-3	154384	152382	66143	154358
	MP	MP	MP	WK	N	SF	F	P	AegL	ArfL
SiO_2	54.60	52.34	49.50	53.30	48.25	49.58	58.50	59.67	52.53	53.06
TiO_2	0.20	0.10	0.28	0.16	0.32	0.29	0.32	0.22	0.23	0.15
ZrO_2	0.75	0.98	1.89	1.30	0.49	0.39	0.27	0.19	1.18	0.74
Al_2O_3	14.53	19.05	19.01	13.40	19.30	18.00	16.21	16.53	13.11	14.29
Fe_2O_3	8.50	6.95	5.29	7.90	4.07	4.68	3.03	3.71	9.83	2.89
FeO	2.40	0.22	1.54	3.42	3.08	4.68	3.80	2.24	3.86	7.75
MnO	0.24	0.26	0.25	0.31	0.21	0.21	0.17	0.16	0.30	0.44
MgO	0.16	0.12	0.13	0.23	0.10	0.14	0.11	0.17	0.09	0.14
CaO	2.78	1.08	1.83	2.10	1.68	1.78	1.76	2.30	1.41	0.38
Na_2O	7.16	9.89	13.27	8.84	14.37	12.85	7.56	7.81	10.63	9.84
K_2O	5.27	4.56	2.88	4.04	3.41	4.00	5.64	4.38	3.05	3.42
Cl	0.03	0.03	0.27	0.04	2.34	1.79	0.12	0.01	0.18	0.02
F	0.88	0.34	0.66	1.05	0.16	0.43	0.20	0.78	0.14	0.15
P_2O_5	0.05	0.09	0.07	0.04	0.06	0.05	0.04	0.03	0.02	0.54
vol/H_2O	2.51	3.56	2.45	3.22	1.44	1.30	1.47	1.60	2.73	4.39
	100.06	99.57	99.28	99.35	99.28	100.17	99.20	99.80	99.29	98.74
- O	0.38	0.15	0.34	0.45	0.61	0.59	0.11	0.33	0.08	0.06
Total	99.68	99.42	98.94	98.90	98.67	99.58	99.09	99.47	99.21	98.68
A.I.	1.2	1.1	1.3	1.4	1.4	1.4	1.1	1.2	1.6	1.4
Cs	10.0	22.2	9.8	5.5	6.2	7.1	5.3	12.2	5.8	7.4
Rb	517	595	514	542	334	395	315	282	435	551
Ba	365	111	78	163	11	8	42	34	110	47
Pb	56	117	61	139	94	59	45	51	312	663
Sr	166	396	109	108	10	9	27	51	58	61
La	403	741	855	743	594	384	244	189	654	3690
Ce	770	1220	1550	1460	1180	889	512	382	1160	5040
Nd	295	493	590	573	540	370	219	153	566	1570
Sm	58.7	81.7	99.5	105	96	54	38.2	23.3	105	176
Eu	5.8	8.1	10.7	7.2	9.6	5.7	3.6	2.6	10.9	17.3
Tb	10.0	15.8	21.2	21	16.1	8.1	5.8	3.8	20.6	23.1
Yb	36.3	47.0	70.2	65.2	38.3	25.1	17.9	11.0	67.3	40.1
Lu	5.0	6.5	9.9	9.3	5.2	3.4	2.4	1.6	8.3	5.3
Y	322	437	717	759	461	192	184	125	679	642
U	12	13	29	23	19.5	14.8	10	8	44	124
Th	31	31	54	67	41	60	28	37	41	63.5
Zr	5549	7250	1.4%	1.2%	4360	2850	2070	1420	8720	5270
Hf	128	186	331	248	76	81	42.5	30.7	181	61
Nb	528	746	1440	894	742	490	325	228	785	499
Ta	34.0	49.8	127	64	52	31	19	10.6	52	39
Zn	227	395	491	619	505	424	276	280	728	2570
Sc	1.4	0.7	0.9	1.5	<0.01	0.2	0.5	0.1	1.0	0.01
Ga	71	99	88	70	73	74	58	66	77	101

104361A, 109303, 104380A, 66143: new analyses; 154314, 154346, 154384, 154358: from Bailey *et al.* (2001); 154350-3 and 152382: J.C. Bailey, personal information.

MP = marginal pegmatite, WK = white kakortokite, N = naujaite, SF = sodalite foyaite, F = foyaite, P = pulaskite, AegL = aegirine lujavrite, ArfL = arfvedsonite lujavrite, A.I. = agpaitic index

Table 6. Selected element ratios for some Ilímaussaq rocks

	104361 A	MP B	109303 MP	104380A MP	104363A MP	107724 m-sc	154346 N	154351 SF	154384 F	154382 P	154314 WK	66143 AegL	154358 ArfL
Na/K	1.2	1.5	1.9	4.1	3.1	1.3	3.8	2.9	1.2	1.6	2.0	3.1	2.6
Na/Ca	2.7	2.7	9.5	7.6	8.4	4.0	8.9	7.5	4.5	3.5	4.4	7.9	27
Na/Zr	9.5	6.2	10.1	7.0	12.6	130	24	33	27	41	5.8	9.0	13.5
K/Rb	85	74	64	78	78	64	85	84	149	129	62	58	53
K/Ba	120	90	341	306	138	202	2573	4150	1114	1069	206	230	604
Ca/Ba	54	49	68	167	51	66	1084	1579	297	480	91	91	57
Ca/Sr	119	70	19	119	116	426	1193	1404	463	320	138	173	44
Ca/Zr	3.6	2.5	1.1	0.9	1.5	32.3	2.7	4.4	6.0	11.5	1.2	2.5	0.002
Rb/Sr	3.1	1.7	1.5	4.7	4.1	18.8	33.4	43.9	11.7	5.5	5.0	7.5	9.0
Ba/Sr	2.1	1.4	0.3	4.7	2.3	6.4	1.0	0.9	1.6	0.7	1.5	1.9	0.8
Ba/Rb	0.7	0.8	0.2	0.15	0.6	0.3	0.03	0.02	0.1	0.1	0.3	0.25	0.1
Zr/Hf	43	41	39	42	49	45	57	35	49	46	48	48	86
Nb/Ta	15.5	13.8	15.0	11.3	19.3	20.3	14.2	15.8	17.1	21.5	14.0	15.1	12.8
Th/U	2.6	3.6	2.4	1.9	2.8	2.9	2.3	4.1	2.8	4.6	2.1	0.9	0.5
La/Y	1.3	1.2	1.7	1.2	1.4	1.2	1.3	2.0	1.3	1.5	1.0	1.0	0.9
Zr/U	462	581	558	483	314	375	223	190	207	178	530	198	42.5
Zr/Th	179	163	234	259	105	104	106	47.5	74	38	179	212	83
Zr/La	13.8	15.1	9.8	16.4	12.9	18.8	7.3	7.4	8.5	7.5	16.1	13.3	1.4
Zr/Y	17.2	17.3	16.6	19.5	17.4	23.2	8.4	14.8	11.3	11.3	15.9	12.8	8.2
Zr/Nb	10.5	9.8	9.7	9.7	10.0	7.1	9.9	5.8	6.4	6.2	13.9	11.1	10.6
Zr/Ba	16.2	20.5	65	179	34	2.1	396	356	49	41	74	79	112
Zr/Sr	33	30	18	128	78	13.2	436	316	77	28	111	150	86

MP = marginal pegmatite, m-sc = mini-scale pegmatite, WK = white kakortokite, N = naujaite, SF = sodalite foyaite, F = foyaite, P = pulaskite, AegL = aegirine lujavrite, ArfL = arfvedsonite lujavrite

consequence of the long interval of consolidation of these rocks from 700-900 to 400-450°C (Sood & Edgar 1970; Markl *et al.* 2001).

It is important to know whether alteration took place in a closed or an open system, i.e. whether the original chemistry is retained or not. The microcline of the matrix rocks forms lucid crystals which have rims and intergrowths of albite but otherwise appear to be little altered. In some samples, nepheline is strongly altered to analcime and sericite which may influence the contents of K_2O, Na_2O and Rb, even if most K and Rb may be retained in sericite. The original amphibole of some matrix rocks is strongly altered and is substituted by aegirine, fluorite and pigmentary material which may influence the contents of Ca, Na, K, Mn and Fe (and Li which is not included in the present study) and the Fe^{2+}/Fe^{3+} ratio. Ferguson (1970a, b) found that agpaitic Ilímaussaq rocks with more than 10% secondary zeolites have lower Na/K ratios than the equivalent rocks with less zeolites and that the Ba/Rb and Ca/Sr ratios show no consistent covariance with zeolitisation. The eudialyte of 104361A, B and 109303 is substituted by zircon, analcime, fluorite, aegirine, ill-defined pigmentary material, and in 109303 additionally a carbonate mineral. Ussing (1894) demonstrated that Zr is retained in these zircon-bearing pseudomorphs after eudialyte. According to Ferguson (1970a), there is no correspondence between the degree of eudialyte alteration or zeolitisation and the Zr/Nb ratio. Andersen *et al.* (1981b) examined the Zr/U and Zr/Y stratigraphy of the kakortokite-lujavrite sequence and found that alteration of the rocks took place in a closed system, i.e. the original element ratios were retained when the mineral carriers of the elements disintegrated. Rose-Hansen & Sørensen (2002) in a study of lujavrites in drill cores found very little variation in the Zr/Y ratio with stratigraphical level irrespective of the degree of alteration of eudialyte, the main carrier of Zr and Y in the rocks, and concluded that Zr and Y were not separated during alteration of eudialyte.

This review of earlier studies shows that the question whether the alteration processes have caused major changes in the chemical compositions of the rocks is still open. Whereas some minor and trace elements and element ratios may be unchanged, loss of alkalis, particularly Na may have occurred. Since rocks representing magma compositions are rare in Ilímaussaq, it is nevertheless worth while examining the information the matrix rocks of the marginal pegmatite may provide about magma compositions and evolution.

In an attempt at evaluating the consequences of the alteration processes, traditional CIPW weight norms were calculated for four of the rocks together with agpaitic norms designed to estimate the primary mineral association of agpaitic rocks. In the traditional CIPW system, peralkalinity is indicated by the normative 'minerals' acmite, *ac*, $NaFeSi_2O_6$, sodium metasilicate, *ns*, Na_2SiO_3, and potassium metasilicate, *ks*, K_2SiO_3. In peralkaline rocks such as the agpaitic Ilímaussaq nepheline syenites, the surplus of alkalis is bound not only in alkali pyroxenes, but also in alkali amphiboles, sodalite, eudialyte and a great number of more or less rare sodium- and potassium-rich minerals. Therefore, two new normative mineral names are introduced, arfvedsonite and eudialyte: *arf* and *eud*:

- *arf*, $Na_3(Fe^{2+})_4Fe^{3+}Si_8O_{22}(OH)_2$ is omitted, maintaining the principle that normative minerals are anhydrous. The potassium content of arfvedsonite is disregarded by analogy with the tradition of considering the normative mineral *ne* as a purely sodic mineral.

- *eud*,
There appears to be no general agreement on the chemical formula of eudialyte. The empirical formula by Johnsen *et. al.*(2001), which is based on a great number of analyses, will be used:

$Na_{15}(Ca,REE)_6(Fe,Mn)_3Zr_3Si_{26}O_{73}(O,OH,H_2O)_3(Cl,F,OH)_2$.

This can be simplified to
$Na_5Ca_2(Fe^{2+},Mn)ZrSi_9O_{25.5}$.

When calculating the agpaitic norm, which for obvious reasons is very approximate, compensation for Cl and F has not been undertaken. Mg, Ti and P norm minerals have not been included in the calculations, because the rocks in question do not contain relevant Ti-, Mg- and P-minerals. P, for instance, is not found in apatite but in REE phosphate minerals such as monazite and britholite, which are not included in the norm. Ti is partially added to Zr to form *eud* in some rocks, but Mg, Ti and P, which are minor components of the rocks, are generally considered to be excess components. Cl and F are either presented as *hl* and *fl*, halite and fluorite, or as excess Cl and F.

The CIPW weight norms of the four rocks (Table 3) clearly fall in two groups: 104361A and 109303 are without *ns* and contain *hm*, hematite, whereas 104363A and 104380A have *ns* and are without *hm*. All four have *ac* and are peralkaline.

The agpaitic norms of the four rocks have *eud* and *ac*, with the exception of 104361A which contains *arf* instead of *ac*. *ns* is missing in the special norms of all four rocks since no Na_2O is left after calculation of *eud*, *arf* and *ab*. The contents of *wo* (wollastonite), *hm* and *fs* (ferrosilite) are explained by the simplified compositions of the norm minerals *eud* and *arf* which cannot account for the range in variation of the chemical compositions of the minerals of these rocks. The differences between the two pairs of rocks may be explained by different degrees of alteration, 104361A and 109303 are altered, whereas 104380A is faintly altered and 104363A is practically unaltered. The low Na_2O contents in 104361A, which is expressed in the lack of *ns* and the presence of *hm*, may indicate a loss of sodium during the alteration of this rock. Alternatively, the low Na_2O content may be a primary feature; it is similar to the contents of Ilímaussaq pulaskite and foyaite (Table 5) and of trachytic-phonolitic rocks intruding the country rocks of the complex (Table 2) (Larsen 1979). The question of the importance of the alteration processes will be further discussed on p. 53.

The chemical analyses of the matrix rocks from the three localities are distinctly different from each other. 109303 from the north coast of Tunulliarfik differs from the others in the highest volatile content, 3.56 weight %, elevated contents of the analcime components Na_2O, Al_2O_3, H_2O and Cs, and an extreme Fe_2O_3/FeO ratio. The high Fe_2O_3/FeO ratios of these rocks are partly due to the presence of primary aegirine, partly due to the alteration of amphibole and eudialyte to aegirine and Fe-rich pigmentary material. The analysis of the fine-grained contact rock (104361B) is similar to that of the adjacent coarser-grained matrix rock (104361A) from the north coast of Kangerluarsuk; the main difference is the higher Al_2O_3 and slightly higher MgO and lower alkalis in the fine-grained rock. This may be partly explained by reactions with the adjacent augite syenite. 104380A from Kvanefjeld differs from the other matrix rocks in bulk composition as well as in element ratios.

The texture and unaltered state of 107724 and 104363A suggest that they have been formed by rapid crystallization of melts and that their compositions have not been modified by secondary processes. Therefore, they may represent the composition of the melts from which they were formed, minus of course the volatile elements, the loss of which may have been the cause of the quenching of the melts.

Relative to matrix rocks 104361A, B and 104363A, the fine-grained black rock from the mini-scale model pegmatite (107724) is enriched in TiO_2, CaO, Na_2O, K_2O, F, Cl and Rb, and highly depleted in Zr and most trace elements which are explained by the abundance of microcline and fluorite and the lack of eudialyte. The major element contents are rather close to the average composition of kakortokite (Ferguson 1970a, b) and to the analysis of white kakortokite of Table 5 but with a lower Fe_2O_3/FeO ratio than kakortokite and, with the exception of Rb and Ba, much lower contents of trace elements. The element ratios Na/Ca, K/Rb, K/Ba, Ba/Rb, Zr/Hf, Th/U, La/Y and Zr/La are, however, similar to the kakortokite ratios, but generally different from the ratios of the marginal pegmatite (Table 6).

The fine-grained rock (107728), which is supposed to represent a quenched pegmatite melt, occurs in interspaces between pegmatite minerals. This and the size of the specimen did not allow preparation of material for chemical analysis. A rough estimate of the chemical composition has been made by calculating the chemical composition of mixtures of 40 wt. % of Na-poor microcline, 30% Al-poor aegirine and 30% eudialyte and of 60% microcline, 20% eudialyte and 20% aegirine. The melt from which the rock crystallized appears to have been relatively enriched in SiO_2 (55-60%), ZrO_2 (2-3%), Na_2O (6-9%), K_2O (5.5-8.5%) and F (0.2-0.3%).

The chemical analysis of the microcline-rich variety of the xenolith in kakortokite (104352) is compared with an analysis of the Narsaq Elv microsyenite (Table 2). The xenolith is richer in total iron and K_2O. Otherwise the two analyses are similar with regard to major elements. 104352 is, however, richer in most trace elements than the microsyenite. 104352 has 1.47% normative *q*, 54.97% *or*, 15.14% *ab*, 0.69% *ns* and 22.74% *ac*.

When compared with the augite syenite, the Ilímaussaq agpaitic rocks are distinctly lower in TiO_2 (< 0.35% versus >1.20%), MgO (<0.25% versus >0.8%), CaO (<2% versus >3%) and P_2O_5 (<0.2% versus >0.3%) and higher in ZrO_2 (>1% versus <0.05%) and Na_2O (>7.5% versus <5.6%), augite syenite data quoted from Bailey *et al.* (2001). The roof and floor series rocks are markedly lower in Ba, Sr, and Sc and higher in most other trace elements than augite syenite. Thus there is a sharp break in chemical composition between augite syenite of stage 1 and the rocks of stage 3.

The role of Ba and Sr

The agpaitic Ilímaussaq rocks have lower contents of Ba and Sr than most or perhaps all other occurrences of agpaitic rocks. The average contents of the nepheline syenites of the complex are 50 ppm Ba and 40 ppm Sr, the Ba/Sr ratio is 1.25 (J. C. Bailey, personal information). The corresponding Ba and Sr values are 700 and 670 ppm for the Lovozero complex and 900 and 680 ppm for the Khibina complex, Kola Peninsula (Gerasimovsky 1969; Gerasimovsky *et al.* 1964/1966). Several Ba and Sr minerals have been found in the two Kola complexes, whereas only three Ba minerals have been identified in pegmatites in the Ilímaussaq complex: barylite, ilimaussite and joaquinite-(Ce). All three are very rare; no Sr minerals have been found.

The average values of Ba and Sr in the Ilímaussaq rocks are: augite syenite 1900 ppm Ba, 284 ppm Sr, Ba/Sr 6.7; pulaskite and foyaite 36 ppm Ba, 25 ppm Sr, Ba/Sr 1.4; sodalite foyaite 12.5 ppm Ba, 11 ppm Sr, Ba/Sr 1.1; naujaite 7 ppm Ba, 11 ppm Sr, Ba/Sr 0.6; kakortokite 159 ppm Ba, 87 ppm Sr, Ba/Sr 1.8; aegirine lujavrite 67 ppm Ba, 94 ppm Sr, Ba/Sr 0.7; and arfvedsonite lujavrite 50 ppm Ba, 61 ppm Sr, Ba/Sr 0.8 (J.C. Bailey, personal information). Thus, the rocks can be divided in three groups on the basis of their Ba and Sr contents: (1) augite syenite, (2) the roof series, (3) kakortokites and lujavrites. The homogeneous matrix rocks of the marginal pegmatite and the quenched pegmatites fall in the third group (Tables 2, 5, 6)

Ba substitutes for K in K-feldspar and mica minerals. The Ba and Sr contents of the microcline of the roof series rocks are low: 7-40 ppm Ba and 5-10 ppm Sr. The microcline of the marginal pegmatite, kakortokite and the lujavrites has higher contents, 100-500 ppm Ba and 10-50 ppm Sr, and up to 362 ppm Sr in the red kakortokite layers (J.C. Bailey, personal information). This can partly explain the elevated contents of Ba and Sr in these rocks.

Sr substitutes for Ca in eudialyte and fluorite, the only minerals with a substantial Ca content in the rocks in question. Eudialyte can also contain Ba (Gerasimovsky 1969; Semenov 1969; Ferguson 1970a). One would therefore expect some degree of covariation between Ca and Sr and between Zr and Sr and Zr and Ba but this is definitely not the case. The two chemically related rocks (104361A and 104363A) have widely different Ca/Zr, Zr/Ba and Zr/Sr ratios, but similar Ca/Sr ratios (Table 6). This may be a result of the alteration of the eudialyte of 104361A. Parts of the naujaite and sodalite foyaite of the roof series have high eudialyte contents but have nevertheless low contents of Sr and Ba and variable Zr/Ba and Zr/Sr ratios. This indicates that eudialyte is not the major carrier of Ba and Sr.

Annotated survey of the literature on the evolution of the Ilímaussaq complex

Since Ussing (1912), all students of the Ilímaussaq complex agree that augite syenite was the first intrusive phase. He concluded that the rocks inside the augite syenite rim form a stratified series, mentioned from the top downward: arfvedsonite granite, quartz syenite, pulaskite, foyaite, sodalite foyaite, naujaite and at the bottom lujavrite with intercalated kakortokite.

The relationship between augite syenite and the agpaitic rocks has evoked discussion. Ussing (1912: 327, fig. 30) argued that an early augite syenite massif was subsequently intruded and almost completely replaced by a nepheline syenite intrusion. The geochemical kinship between augite syenite and nepheline syenites made Sørensen (1958), Ferguson (1964, 1970a, b), Gerasimovsky (1969) and Engell (1973) propose a closed system origin for the Ilímaussaq complex. The initial magma was represented by the chilled silica-saturated augite syenite (Hamilton 1964). As crystallization proceeded, the magma became undersaturated and increasingly peralkaline which led to the formation of the agpaitic rocks inside the shell of augite syenite. The recognition of mineralogical and chemical discontinuities between augite syenite and the agpaitic rocks and advances in the knowledge of the field relations of the rocks made Sørensen (1970, 1978), Larsen (1976), Nielsen & Steenfelt (1979), Steenfelt (1981) and Larsen & Sørensen (1987) support Ussing's scheme of successive intrusions. Now, at least three main stages of intrusion are recognised: (1) augite syenite, (2) alkali granite, (3) nepheline syenites.

The augite syenite

North of Tunulliarfik, augite syenite is observed along parts of the west contact and in the roof zone of the complex. Augite syenite xenoliths are found in lujavrites at Kvanefjeld, near the west contact of the complex in the Narsaq Elv bed and on the north coast of Tunulliarfik. South of Tunulliarfik, there is a continuous augite syenite rim along the west, south and south-east contacts, from Tunulliarfik to close to Appat. Additionally, there are augite syenite xenoliths in kakortokite and lujavrite; major bodies occur at about 600 m a.s.l. at Laksefjeld (Figs. 1, 20, 21). Small xenoliths occur in pulaskite (Fig. 22).

The seven 200 m deep bore holes, which are distributed over the complex (Fig. 1), do not contain any traces of augite syenite. It appears that augite syenite is confined to the the south and west contact, Kvanefjeld (500-600 m a.s.l.) and the Narsaq Elv bed (300 to 100 m a.s.l.). At Nakkaalaaq (1100-600 m a.s.l.) augite syenite is intruded by granite and quartz syenite sheets.

Minor syenite intrusions

Occurrences of alkali syenite are confined to the same part of the Kvanefjeld plateau as augite syenite and to the Narsaq Elv valley immediately below Kvanefjeld (not marked on Fig. 1). This syenite is a greenish to bluish, fine- to

Fig. 20. Rafts of augite syenite enclosed in marginal pegmatite immediately inside the contact against augite syenite which is seen on the left hand side. Laksetværelv at 430 m a.s.l. Note the fracturing in the augite syenite.

medium-grained, massive rock which is found as xenoliths in arfvedsonite lujavrite and as at least three generations of sheets intersecting xenoliths of augite syenite and the volcanic rocks of the roof. The syenite consists of microcline, albite, arfvedsonite, aegirine and minor neptunite and pectolite. Nepheline, sodalite and analcime may have been introduced from the enclosing lujavrite (Nielsen 1967; Sørensen *et al.* 1969; Nielsen & Steenfelt 1979). The syenite contains anorthosite xenoliths and plagioclase xenocrysts. The alkali syenite is clearly younger than augite syenite and older than arfvedsonite lujavrite; its relations to naujaite and alkali granite cannot be decided because of lack of contacts.

Dykes and sheets of aphanitic to fine-grained peralkaline microsyenite intrude kakortokite and aegirine lujavrite

Fig. 21. The west contact of the complex, south coast Kangerluarsuk. Background: granitic basement, the highest moutain is Iviangiusaq (903 m). The contact between reddish-brown crumbling augite syenite and granite is sharp. Extreme left: the marginal pegmatite. Centre right at coast: white blocks of quartzitic sandstone enclosed in vegetation covered augite syenite.

in the southern part of the complex and naujaite in the bed of Narsaq Elv (Rose-Hansen & Sørensen 2001). They are marked as volcanic rocks on the map of Ferguson (1964). Apart from finer grain size, these rocks are practically identical with the above-mentioned alkali syenite. A specific feature is that both have abundant neptunite and pectolite. The microsyenites have distinctly intrusive relationships to naujaite and to aegirine lujavrite and kakortokite. The relationship to arfvedsonite lujavrite is ambiguous due to poor exposures of contacts. Most features indicate that arfvedsonite lujavrite intrudes the microsyenite, i.e. the same relationship as between alkali syenite and arfvedsonite lujavrite.

Chemical analyses of the Kvanefjeld alkali syenite and the microsyenites are published by respectively Sørensen *et al.* (1969) and Rose-Hansen & Sørensen (2001). They have 61-64% SiO_2, 14-15% Al_2O_3, 3-5% Fe_2O_3, 0.5-2.5% FeO, 0.5-2% CaO, 7-11% Na_2O and 0.2-6% K_2O, 0.3-3 *ns* and 7-16 *ac*. The variations in Na_2O and K_2O and in Fe^{3+} and Fe^{2+} are due to varying proportions of respectively microcline-albite and arfvedsonite-aegirine; a considerable part of this variation is a result of secondary processes.

Most samples of alkali syenite and the microsyenite sheets are *q* normative which excludes a direct genetic association with the agpaitic rocks.

The granite and quartz syenite

Ussing (1912) considered the sequence quartz syenite, pulaskite and foyaite as a transition zone between the overlying granite and the underlying agpaitic rocks, although the pulaskite could be a remnant of the syenite that at one time filled the reservoir and, in that case, only the quartz syenite and foyaite were transitional rocks. With regard to the granite he wrote: "It is perhaps not impossible to imagine that the granite in some way may have originated from the syenitic magma. Owing to its lower specific gravity it may have accumulated at the top of the chamber, while both magmas were yet in a semi-fluid condition" (Ussing 1912: 339). He based this interpretation on the observation that sandstone xenoliths in augite syenite on the south coast of Kangerluarsuk (Fig. 21) have reacted with the syenitic magma under formation of alkali granite. Sørensen (1958, 1966, 1970) noted that granite and quartz syenite only occur in the uppermost part of the complex and therefore may have been formed by assimilation of acid roof rocks in the augite syenitic magma. Ferguson (1964) considered the granite to be intrusive into the foyaite and regarded the quartz syenite and pulaskite as hybrids formed by reaction between granitic and agpaitic melts. Hamilton (1964) thought that the granite was the youngest member of the complex. Steenfelt (1981) has, however, demonstrated that granite and quartz syenite belong to a separate intrusive phase which postdates augite syenite and predates the agpaitic rocks. This interpretation is in excellent accord with the occurrence of independent intrusions of alkali-acid rocks elsewhere in the Gardar province; the closest example is the Dyrnæs-Narsaq complex immediately to the west of the Ilímaussaq complex (Upton *et al.* 2003).

The nepheline syenites

These rocks are divided into a roof series, an intermediate series and a floor series. The prevailing view is that they are comagmatic and formed in a closed system (Ferguson 1964, 1970b; Larsen & Sørensen 1987; Markl *et al.* 2001; Marks *et al.* 2004).

The roof series: pulaskite, foyaite, sodalite foyaite and naujaite

Ussing (1912: 341) was uncertain about

the relationship between augite syenite and pulaskite in the uppermost part of the nepheline syenites since these two rocks are in contact on the NE side of Nunasarnaasaq (Fig. 1) and no boundary between them could be observed. He pointed out that pulaskite could either be a product of reaction between granitic and agpaitic melts or have been formed by consolidation of the syenitic melt that at one time filled the magma chamber. Sørensen (1958) accepted the last-named view and thought that pulaskite represents the primary magma of the agpaitic part of the complex. Ferguson (1964) supported Ussing's first-named option that pulaskite is a reaction product between granite and agpaites since there is a gradual transition from pulaskite to the overlying quartz syenite. The discovery of angular xenoliths of augite syenite in pulaskite on the NE side of Nunasarnaasaq (Fig. 22) proves that augite syenite consolidated before the emplacement of the pulaskite (Sørensen 1978). Nielsen & Steenfelt (1979) and Steenfelt (1981) produced substantial evidence that pulaskite is a true magmatic rock and it is now generally accepted that pulaskite is the uppermost and earliest member of the roof series of the complex.

Hamilton (1964) noted that a pulaskite-looking rock occurs in the east contact of the complex between Nakkaalaak and Tunulliarfik. This has to be further investigated. In the northern part of the complex, naujaite forms the major part of the marginal contact against the volcanic country rocks, lujavrites a minor part.

Ferguson (1964) and Hamilton (1964) considered foyaite (heterogeneous syenite of Ferguson 1964) as an internal differentiate of the augite syenite. However, the local gradual transition from pulaskite to the underlying foyaite makes foyaite a member of the roof series. In other places pulaskite occurs as xenoliths in foyaite (Steenfelt 1981).

Ussing (1912) thought that sodalite foyaite, the chemical composition of which is very close to the average composition of the agpaitic rocks of the complex (Table 4), was formed by consolidation of undifferentiated agpaitic magma intruded into a shrinkage fracture on top of the cooling naujaite. There is, however, a gradual transition from foyaite to sodalite foyaite in some places, and in other places sodalite foyaite forms patches within foyaite (Steenfelt 1981). Thus, a continuous evolution from pulaskite via foyaite to sodalite foyaite has been demonstrated but with irregularities in the form of lenticular bodies and screens of overlying rocks enclosed in the underlying rocks.

The transition from sodalite foyaite to the underlying naujaite is gradual but in some places, the transition zone shows an alternation of horizons of naujaite and sodalite foyaite. In the outlier of roof series rocks south of Taseq (Fig. 1), the transition from pulaskite via foyaite to sodalite foyaite is gradual, but the contact between naujaite and pulaskite is sharp, the naujaite showing decreasing grain size towards the contact (Steenfelt 1981: 48).

Pulaskite is only found near the roof of the agpaitic magma chamber. If pulaskite crystallized from the initial magma of the roof series rocks, one would expect to find pulaskite also along the lateral contacts which, with the exception of the contact on the NE side of Nunasarnaasaq, is not the case. This dilemma will be considered on p. 49.

The formation of naujaite

Ussing (1912) concluded that naujaite was formed by flotation accumulation of sodalite crystals in the upper part of the magma chamber, a view accepted by all subsequent investigators of the complex.

Sodalite contents of naujaite varying from 20-30 to more than 75 vol. % indi-

Fig. 22. Angular xenoliths of augite syenite (brownish) in pulaskite. East side of Nunarsarnaasaq.

cate that the agpaitic melt at the naujaite stage was extremely enriched in Na, Cl and S. The vertical distance between sodalite foyaite and lujavrite, that is the thickness of the main body of naujaite, is at least 600 m (Andersen *et al.* 1981a; Rose-Hansen & Sørensen 2002). But originally, the naujaite zone was considerably thicker which is shown by several horizons of naujaite rafts in the underlying arfvedsonite lujavrite and aegirine lujavrite. This zone of naujaite rafts in lujavrite was called the breccia zone by Ussing (1912). It is found on both sides of Tunulliarfik. The north wall of Tunulliarfik presents a vertical section (Fig. 23), the plateau between Tunulliarfik and Kangerluarsuk an approximate horizontal section through the breccia zone. Additional evidence of a former greater thickness of the naujaite is provided by the naujaite masses in the marginal pegmatite located several hundred metres below the present main naujaite horizon. The exposed naujaite rafts in lujavrite are up to about 20 m thick (Fig. 23). However, bore hole II on the north coast of Tunulliarfik ends in 80 m naujaite (from 33 to 118 m below sea level), and bore hole VI in the southern part of the complex has naujaite from *c.* 230 m to 150 m a.s.l., the end of the core. In bore hole II, naujaite is overlain by about 120 m arfvedsonite lujavrite and minor aegirine lujavrite and furthermore by more than 100 m lujavrite in the mountain above the site of the bore hole. In bore hole VI, the naujaite is overlain by 120 m of mainly arfvedsonite lujavrite. This bore hole is located on a plateau and the original thickness of the overlying lujavite is unknown. Thus, the large naujaite masses in bore holes II and VI are separated from the overlying main mass of naujaite by at least 120 m of lujavrite, i.e. lujavrites are sandwiched between large masses of naujaite. The neighbouring bore holes do not intersect naujaite in these depth intervals which means that the naujaite bodies have a limited horizontal extent. These naujaite masses may be sunken

Fig. 23. The north wall of Tunulliarfik. Left hand side, the west contact of the Ilímaussaq complex against Gardar basalts (BA). In the central part of the photo, the lower part of the mountain wall consists of arfvedsonite lujavrite (black) with rafts of naujaite (white). It is overlain by naujaite and sodalite foyaite. The highest tops (BA) on the right side consist of the supracrustal rocks that form the roof of the intrusion. Tuttup Attakoorfia is the small black coast cliff left of the iceberg (centre right).

blocks from the higher-lying main zone of naujaite or be screens which are remnants of deeper-lying naujaite left more or less *in situ* when the lower part of the original naujaite body was intruded by lujavrites.

There are good reasons to believe that the main mass of naujaite originally persisted across the whole complex corresponding to an area of about 100 km^2. The volume of a 600 m thick naujaite mass is about 60 km^3. Of that volume, 20 to more than 45 km^3 are occupied by sodalite. To that should be added the sodalite of the naujaite rafts in the underlying lujavrite. Such a concentration of sodalite would require crystallization of sodalite in an enormous magma reservoir, which, however, is difficult to match with the known dimensions of the agpaitic magma chamber. The thickness of the agpaitic sequence, from sodalite foyaite near the roof of the complex to the lowermost exposed kakortokite, is about 1500 m (Andersen *et al.* 1981a). Heat flow measurements (Sass *et al.* 1972) indicate that the exposed agpaitic rocks are underlain by at most 1-2 km of rocks containing appreciable amounts of U, Th and K. Thus, the magma chamber was shallow. It was probably located on top of the cumulates formed simultaneously with pulaskite to sodalite foyaite and perhaps naujaite which can be estimated to have low contents of U and Th. Gravity data indicate that the top of the heavy rocks underlying the Ilímaussaq region is located close to the surface (Forsberg & Rasmussen 1978; Blundell 1978; Upton & Blundell 1978). Thus, the height of the disc-shaped agpaitic magma chamber from the roof to the hidden non-agpaitic cumulates was from 2500 to 3500 m and the total volume 250 to 350 km^3. It may be estimated that more than five and perhaps more than ten per cent of the entire volume of agpaitic rocks is made up of sodalite.

The enormous quantity of sodalite components needed to form naujaite may be partly explained by successive accumulation of residual liquids below a downward moving crystallization front in a sealed magma chamber (Larsen & Sørensen 1987). Periodical changes in conditions of crystallization could arrest crystallization and form pegmatitic horizons in foyaite and naujaite and recurrence of horizons of sodalite foyaite in naujaite (Ferguson 1964; Steenfelt 1981; Sørensen & Larsen 1987).

The occurrence of naujaite and sodalite-rich pegmatites along the contact between augite syenite and kakortokite to the south of Kangerluarsuk show that sodalite was formed below what is now the lowermost naujaitic part of the roof series. Therefore, sodalite most probably crystallized over a considerable depth interval in the magma chamber. Crystals were plastered to the lateral walls and accumulated on the floor of the chamber and clouds of sodalite crystals floated in the chamber.

This scheme of formation of naujaite is mirrored by the above-mentioned pegmatitic mini-scale model of the agpaitic magma chamber. The apical sodalite-rich rock comprises 10 cm of the height of the cupola of 25 cm, corresponding to the accumulation of sodalite in naujaite. The size of the cupola

Fig. 24. The Lakseelv valley. Right-hand side: Laksefjeld (LF, 680 m), the lower part of which is layered kakortokite, the upper part aegirine lujavrite (dark). The light-coloured rock in the upper part is a large raft of augite syenite (see Fig. 1). Left-hand side, the Lakseelv Valley, in background, the lowermost lujavrites with naujaite xenoliths. Behind Laksefjeld, the basement granite in the mountain Killavat (the comb) of which the highest peak is 1216 m.

was insufficient for the formation of interstitial minerals in the sodalite-rich upper part and layering in the lower part. Clusters of sodalite crystals in the lower part of the pegmatite cupola show that sodalite crystallized over a depth interval and that the crystals clustered together but did not always rise to the apical part before consolidation of the melt prevented further movement. The fact that the sodalite-rich rock tapers out downward along the lateral contacts of the pegmatite cupola is also indication of sodalite crystallization over a depth interval corresponding to the occurrence of naujaite in the marginal pegmatite in the south-west contact of the complex.

According to fluid inclusion studies, sodalite began crystallization at a pressure of 3.5 kbars corresponding to a depth of 10 km, whereas the roof series rocks are estimated to have formed at pressures around 1 kbar (Markl *et al.* 2001). This indicates that sodalite crystallized already during the ascent of the magma from the deep-seated parental magma chamber.

The kakortokites and lujavrites

Ussing (1912) followed by Sørensen (1958) considered the kakortokite to be an intercalation in the lujavrites and regarded lujavrite-looking rocks located at the mouth of Lakseelv as lujavrites underlying the main series of kakortokite. Ferguson (1970a) showed that these 'hybrid' rocks are related to kakortokite and Bohse *et al.* (1971) that they are layered kakortokite deformed and mixed by slumping processes, i.e. the kakortokites are the lowermost exposed rocks of the complex. Exposures at Laksefjeld (Fig. 24) and in the Lakseelv valley show a gradual transition from kakortokite into the overlying aegirine lujavrite. This is seen as a decrease in grain-size, aegirine substituting arfvedsonite as the dominating mafic mineral, separate grains of microcline and albite instead of perthitic alkali feldspar and partly different minor and accessory minerals. The aegirine lujavrite is divided into a lower aegirine lujavrite I and an upper aegirine lujavrite II. (Bohse *et al.* 1971; Andersen *et al.* 1981a; Bohse & Andersen 1981). A zone of alternating layers of aegirine lujavrite and arfvedsonite lujavrite marks the transition from aegirine lujavrite to the overlying arfvedsonite lujavrite. The same succession of aegirine lujavrite and arfvedsonite lujavrite has now also been found near the east contact of the complex on the north coast of Tunulliarfik, an indication that the lujavrite stratigraphy established in the Laksefjeld-Lakseelv area can be extended to the part of the complex lying to the north of Tunulliarfik. This interpretation is supported by the logs of drill cores III and IV from respectively the north and south coasts of Tunulliarfik (Fig. 1, and Rose-Hansen & Sørensen 2002). The arfvedsonite lujavrite intruded the overlying naujaite during formation of what Ussing (1912) called the breccia zone and also formed dykes and sheets in the naujaite.

The arfvedsonite lujavrite passes locally into hyperagpaitic naujakasite lujavrite in which nepheline is substituted by naujakasite and eudialyte by steenstrupine (Khomyakov *et al.* 2001; Sørensen & Larsen 2001; Andersen & Sørensen 2005).

Various types of layering in kakortokites and lujavrites are described and discussed by Ussing (1912), Sørensen (1969), Sørensen & Larsen (1987) and Bailey *et al.* (2006). Unique spheroidal structures in arfvedsonite lujavrite are interpreted as products of a late-stage, low-temperature immiscible separation of a water-rich melt fraction in a Na-rich lujavritic melt (Sørensen *et al.* 2003).

The exposed part of the kakortokite series and the lujavrite series are estimated to be respectively about 300 m and more than 500 m thick (Andersen *et al.* 1981a; Rose-Hansen & Sørensen 2002).

According to Ussing (1912), the lujavrites were formed by fractional crystallization of the naujaitic magma. Ferguson (1964, 1970a), Gerasimovsky (1969), Sørensen (1970) and Engell (1973) proposed that naujaite formed in the upper part of a closed magma chamber simultaneously with the accumulation of the layered kakortokite in the lower part of the chamber, and that the lujavrite sequence formed by consolidation of the residual magma that was sandwiched between the downward growing naujaitic roof and the upward growing kakortokite. The simultaneousness of naujaite and kakortokite has been quetioned. The amphibole of the kakortokite is higher in Mg and lower in Ca than that of the naujaite (Larsen 1976). Furthermore, the lowermost naujaite is definitely older than the kakortokite since naujaite xenoliths in kakortokite can be demonstrated to have been derived from the lowermost part of the naujaite (Steenfelt & Bohse 1975). This suggests that consolidation of the agpaitic melts was discontinuous. The occurrence of sodalite in the lowermost exposed kakortokites may indicate an overlap from naujaite to kakortokite (Larsen & Sørensen 1987). These problems will be further discussed in a later section.

Lamprophyre dykes

A small number of thin lamprophyre dykes intersect all rocks and are much younger than and not related to the Ilímaussaq magmatism. A few chemical analyses are presented by Sørensen *et al.* (1969).

Derivation of the agpaitic melts

Larsen & Sørensen (1987: 476) proposed that the augite syenitic and the nepheline syenitic melts that formed the Ilímaussaq complex originated in a deep-seated compositionally layered alkali basaltic magma chamber of considerable vertical extent. In this chamber, the residual melts tended to remain physically separate and to collect in its upper part. According to this model, augite syenite and the agpaitic rocks were formed successively by fractionation processes in the magma chamber. More or less continuous trends in parameters such as the activity of SiO_2, NaCl and H_2O support the model of successive batches of continuously fractionating magma (Markl *et al.* 2001). The interpretation is supported by studies of Nd isotopes (Stevenson *et al.* 1997) and of Nd and O isotopes (Marks *et al.* 2004) according to which the agpaitic rocks were formed from a fractionating mantle-derived isotopically homogenous magma.

The occurrence of anorthosite xenoliths in various rocks of the Gardar province, including Ilímaussaq augite syenite and alkali syenite, is taken as evidence of fractionation of plagioclase in the basaltic magma chamber which resulted in an extraction of Al and Ca and in enrichment in alkalis in the magma (Bridgwater & Harry 1968; Upton 1996; Markl *et al.* 2001; Upton *et al.* 2003). The geophysically indicated heavy rocks beneath the Ilímaussaq complex (Forsberg & Ramussen 1978; Blundell 1978) may also have been formed by mineral fractionation processes in the magma chamber.

Trachytic-phonolitic dykes occurring in the country rocks of the complex prove the existence of peralkaline undersaturated magmas (Larsen & Steenfelt 1974; Bailey *et al.* 2001; Sørensen *et al.* 2006). They may have been derived from the complex, or the Ilímaussaq magmas and the magma forming the dykes may have a common origin (Larsen 1979; Marks & Markl 2003). Chemical analyses of dykes are listed in Table 2.

Evidence of the formation of agpaitic melts is provided by ejecta from shallow

Fig. 25. Detail of Fig. 23, the coastal cliff at Tuttup Attakoorfia showing rafts of naujaite enclosed in layered arfvedsonite lujavrite. The lujavrite layers drape over and are conformable with the upper surface of the naujaite rafts and with their internal structures.

Fig. 26. Inch-scale layering in pulaskite, south of Tupersuatsiaaq.

phonolitic and trachytic magma chambers on Tenerife (Wolff & Toney 1993) and the Agua de Pau volcano, São Miguel, Azores Islands (Ridolfi *et al.* 2003). The nepheline syenitic and syenitic ejecta from these volcanoes have interstitial glass and interstital minerals showing agpaitic compositions and mineralogy. The Agua de Pau volcano has produced weakly peralkaline undersaturated and oversaturated syenitic ejecta. It is inferred that the syenites were derived from solidification zones around the margins of the trachytic magma chamber, the undersaturated syenites from the side walls, and the oversaturated syenites, which are enriched in incompatible elements relative to the undersaturated rocks, from the roof.

Similar to the Tenerife and Azores Islands examples, the initial agpaitic Ilímaussaq magmas were born in a crustal magma chamber at a time when the parent magma had acquired a syenitic composition (Bailey *et al.* 2001; Sørensen 2003). Fractional crystallization of augite syenitic melts in a closed system can yield highly reduced, strongly alkaline silica-undersaturated alkaline melts (Marks & Markl 2001). Engell (1973) calculated that 90 to 95% of the initial augite syenite intrusion had solidified before the sodalite foyaite stage was reached. This would demand a 15-36 km deep magma chamber. Larsen & Sørensen (1987) found that the formation of the agpaitic magma by conventional fractionation processes demanded 98% consolidation of the parental basalt magma, and Bailey *et al.* (2001) considered an augite syenitic melt to be the immediate parent of the nepheline syenites and that the final lujavrites formed after 99% crystallization of the augite syenitic magma.

Crystallization of the agpaitic magma

In the sandwich model, the agpaitic Ilímaussaq rocks formed by consolidation of one batch of magma under an impermeable roof that prevented residual melts and volatile components from escaping (Larsen & Sørensen 1987; Markl *et al.* 2001).

The roof series crystallized from the top downward. Pulaskite, the first rock to form, crystallized from a weakly peralkaline and slightly silica-undersaturated melt. The liquidus assemblage was alkali feldspar, fayalitic olivine, clinopyroxene, titanomagnetite and apatite (Ussing, 1912; Sørensen, 1968; Engell, 1973; Larsen 1976, 1977). A part of these minerals were plastered to the roof and walls of the magma chamber, in places in the form of inch-scale type layering (Fig. 26), but the bulk of the early minerals must have sunk to deeper levels in the magma chamber to form the hidden floor series.

Sodalite is an important mineral in the sodalite foyaite and became the liquidus phase in the naujaite because of continued fractionation and accumulation of volatile components in the magma. In melting experiments on naujaite at a water vapour pressure of 1030 bars, sodalite was the liquidus phase at 910°C and the solidus temperature was 430°C, an interval of consolidation of almost 500°C (Sood & Edgar 1970). The sodalite crystals floated in the volatile-rich magma and were poikilitically enclosed by large crystals of microcline, aegirine, arfvedsonite, aenigmatite and eudialyte resulting in the formation of naujaite.

Peralkaline melts have a high capacity for dissolving volatile components (Kogarko 1977). Fractionation of the early water- and volatile-free and Na-free or Na-poor minerals resulted therefore in the formation of increasingly Na- and volatile-rich melts. The amount of anhydrous, Na-free mafic liquidus minerals decreased downward to become subordinate in sodalite foyaite, very rare in naujaite and not observed in lujavrites and kakortokites. Nepheline, amphibole and aenigmatite were interstitial components in the early roof series rocks but increased in amount downward and became major minerals in foyaite and sodalite foyaite. The last-named rock marks the first appearance of agpaicity. The primary mafic minerals were substituted by alkali amphiboles, alkali pyroxenes, aenigmatite and eudialyte (Larsen 1976, 1977; Markl *et al.* 2001). The same order of crystallization is observed in an agpaitic dyke south of the complex (Marks & Markl 2003). An analysis of the intensive crystallization parameters of the agpaitic Ilímaussaq rocks is presented by Markl *et al.* (2001).

In the model of Larsen & Sørensen (1987), kakortokite of the floor series was formed on the floor of the magma chamber by successive upward crystallization of a layered magma. The magma was at this stage nearly volatile saturated, and each layer crystallized in response to the upward loss of a certain amount of volatiles. There is a gradual transition from kakortokite to the overlying aegirine-rich lujavrite.

Structural relations and mechanism of emplacement of the complex

The major part of the surface area of the Ilímaussaq complex is constituted by naujaite, lujavrite and kakortokite (Fig. 1). Pulaskite, granite and augite syenite form the surface exposure of the complex at Nakkaalaak, and pulaskite, foyaite and sodalite foyaite are exposed at the surface in a few small areas north and south of Tunulliarfik.

The complex is divided into a northern and a southern part by a fault through the Lakseelv valley and Kangerluarsuk (Fig. 1). South of the fault, the lowermost exposed part of the complex, the kakortokites, occupy most of the exposed volume. Individual layers can be followed right across the complex from contact to contact. Upwards, the kakortokites grade into aegirine lujavrites which are found on both sides of the fault. On the north side of the fault, the aegirine lujavrites are on both sides of Tunulliarfik succeeded by arfvedsonite lujavrites which intrude the naujaite under formation of the breccia zone. The breccia zone and the overlying naujaite and sodalite foyaite occupy most of the surface area of the agpaitic part of the complex north of the fault.

Faults

The Kangerluarsuk and Tunulliarfik fjords mark the location of two prominent ENE fault zones which were active before the emplacement of the Ilímaussaq complex. The Kangerluarsuk-Lakseelv fault zone was reactivated after formation of aegirine lujavrite and perhaps also arfvedsonite lujavrite, whereas the Tunulliarfik fault remained inactive during and after the emplacement of the agpaitic suite.

The Kangerluarsuk-Lakseelv fault is a hinge fault. The displacement is negligible in its north end at Appat (Fig. 1). At the mouth of Lakseelv, the side to the north of the fault is down-faulted about 350 m compared to the south side of the fault (Bohse *et al.* 1971). At the west contact of the complex about four km farther to the south-west, the lowermost exposed part of the marginal pegmatite-kakortokite series occurs on the south coast of Kangerluarsuk. On the opposite side of the fjord, marginal pegmatite, aegirine lujavrite I and naujaite occur in the coast. The fact that arfvedsonite lujavrite is not found here shows that the lujavrite series wedges out towards the west. The exposed kakortokite series is about 285 m thick. Aegirine lujavite I, the lowermost part of the lujavrite series, is about 80 m thick (Andersen *et al.* 1981a). The contact between this rock and the underlying kakortokite is located 300-400 m above sea level at Laksefjeld in the south-eastern part of the complex (Figs. 1, 24). At the west contact this gives a down-faulting of at least 500 m of the block lying to the north of the fault.

South of Kangerluarsuk, the Ivianguisaq mountains immediately west of the contact of the complex are about 900 m high and the Killavat mountain ridge south of the complex more than 1200 m. These mountains consist of the basement granite and there is no overlying sandstone, but sandstone xenoliths in augite syenite close to the west contact at the south coast of Kangerluarsuk (Fig. 21) show that there must have been a cover of sandstone when the augite syenite was emplaced. North of Kangerluarsuk, the basement granite is overlain by sandstone at an elevation of about 300 m, i.e. this side of the fjord has been down-faulted at least 600 m relative to the south side during the successive pre-agpaitic and post-aegirine lujavritic faulting events. From the north coast of Kangerluarsuk to the south coast of Tunulliarfik, the contact between granite and sandstone falls from about 300 m to sea level. Aegirine lujavrite I occurs in the coast on the north coast of Kangerluarsuk, aegirine lujavrite II on the south coast of Tunulliarfik (Andersen *et al.* 1988). This may be a result of tilting of the landmass between the two fjords. On the north side of Kangerluarsuk, basaltic rocks overlie the sandstone at about 700 m elevation, whereas north of Tunulliarfik, basaltic lavas occur in the coast. This gives a pre-Ilímaussaq downthrow of the north side of Tunulliarfik of at least 700 m.

The ENE direction of the two fault zones is also the strike direction of a regional dyke swarm including the Tugtutôq giant dykes (Upton *et al.* 2003).

The Dyrnæs-Narssaq complex west of the Ilímaussaq complex is intersected by an EW-WNW fault which, however, is cut by the Ilímaussaq complex.

Emplacement

The Ilímaussaq complex is located in prolongation of the giant dykes at the intersection of the ENE fault zone and the EW-WNW fault. The longest axis of the ellipsoid-shaped Ilímaussaq complex is perpendicular to this strike direction (cf. Stephenson 1976). Furthermore, the location of the complex may be related to the discontinuity between the basement granite and the Gardar supracrustals. The complex is in contact with basement granite south of the Lakseelv-Kangerluarsuk fault and with basement granite, sandstone and basaltic rocks north of the fault. On the south coast of Tunulliarfik it is in contact with sandstone and basalts and on the north coast with basaltic lavas in the west contact and with sandstone overlain by basaltic lavas in the east contact.

During the nepheline syenite stage of intrusion, the magma must have occupied a chamber measuring 8 x 17 km and more than 2.5 km from the roof of volcanic rocks to the top of the unexposed floor cumulates that are considered to have been formed simultaneously with the roof series rocks. The space problem was examined by Ussing (1912: 292) who concluded that neither filling of a cavity formed in connection with a major volcanic eruption nor forceful intrusion were supported by the observed field relations. Instead he preferred the then new hypothesis of overhead stoping (Daly 1903), possibly combined with assimilation processes.

Ussing mentioned two examples of stoping: the sandstone blocks in augite syenite on the south coast of Kangerluarsuk and blocks of basalt and naujaite in lujavrite at Nunasarnaq. Additional examples are xenoliths of basalt and augite syenite in lujavrite near the west contact on the north coast of Tunulliarfik, augite syenite xenoliths in marginal pegmatite and kakortokite, naujaite rafts in lujavrites and a variety of xenoliths in lujavrites at Kvanefjeld in the roof zone of the complex. Naujaite encloses rare augite syenite xenoliths close to the contacts. Ussing reported one example of assimilation: the occurrence of quartz in lujavrite in contact with sandstone in the east contact of the complex.

Two outliers of the volcanic cover of the magma chamber have been spared by erosion and cap the two peaks (1270 and 1185 m) of Nakkaalaaq (Fig. 1). This means that a vertical section of about 1500 m through the complex is exposed.

The external contacts of the agpaitic part of the complex are generally smooth and sharp but are intersected by nepheline syenite apophyses and lujavrites protrude into the country rocks in a number of places. The augite syenite has developed chilled contact zones against the country rocks and locally the marginal pegmatite and naujaite display fine-grained contact zones. Apart from that, chilled contacts are rare in the rocks of stage 3.

At the east side of the Ivianguisaq mountains, Ussing (1912: 55) described a small intrusive ellipsoidal arfvedsonite lujavrite body surrounded by the basement granite at 550 m a.s.l. This occurrence has, as far as I know, not been rediscovered.

Large xenoliths of naujaite and basalt occur in a pell-mell inside the east contact at Nunasarnaq (Fig. 1). Some xenoliths consist of both rock types. At this locality, the marginal pegmatite and lujavrites form the contact with the supracrustal country rocks only at a low level in the complex. At higher elevations, naujaite is the contact rock of the complex. The xenoliths consisting of both naujaite and basalt may therefore be large masses of contact rocks that have been detached at higher levels and tumbled into the magma chamber, depressing and bending the partly consolidated lujavrite layers on the magma floor. In this context it should be noted that the adjacent country rocks consisting of sandstone and overlying volcanic rocks show a gentle dip towards the intrusion indicating that the walls of the magma chamber gave in to the weight of the country rocks which locally resulted in the detachment of the mentioned large pieces of the contact assemblage measuring several hundred metres across. Much smaller xenoliths of augite syenite, volcanic rocks and naujaite are enclosed in lujavrites at the west contact of the complex exposed on the north coast of Tunulliarfik. This melange of rocks bears evidence of intense shearing and deformation.

At the south and the east contacts, prominent vertical fractures are developed in augite syenite (Fig. 20) and in the immediate country rocks (Fig. 6). The fracturing has detached rafts of augite syenite and country rocks which fell into the magma chamber and thus facilitated the emplacement of the agpaitic rocks. At the Kvanefjeld plateau, xenoliths of volcanic rocks, gabbro, anorthosite, augite syenite and alkali syenite in arfvedsonite lujavrite are strongly sheared and even folded (Sørensen *et al.* 1969). In the scree slope east of Kvanefjeld and in the foothill of Ilimmaasaq mountain, vertically sheared volcanic rocks from the roof are found together with pulaskite and lujavrite. The scree cover prevents a closer study of the interrelations of these rocks but it appears that also here the country rocks are fractured and sheared close to the contact.

The emplacement of the kakortokite-lujavrite sequence will be discussed in a later section.

The structural position of the roof series

Pulaskite, alkali granite and augite syenite form the surface exposure in the Nakkaalaak-Taseq area (Fig. 1) and xenoliths of pulaskite, foyaite and naujaite occur in lujavrites on the Kvanefjeld plateau and immediately to the east of the plateau in lujavrites below the contact on the basalts in the foothill of the Ilimmaasaq mountain (Sørensen *et al.* 1969, 1974; Rose-Hansen *et al.* 1977). Foyaite and naujaite xenoliths occur in kakortokite in the southern part of the complex (Bohse et al. *1971*; Andersen *et al.* 1988). This suggests that originally the roof series rocks formed the top part of the whole complex in the way it is now seen in the Nakkaalaaq-Taseq ridge and plateau (Fig. 1) and that the naujaite is underlain by lujavrites and these again by kakortokites throughout the whole complex.

The occurrences of pulaskite, foyaite and sodalite foyaite are located at a considerable range of altitudes across the complex. This variation is illustrated by data for pulaskite. From north to south, the few metres thick zone of pulaskite is found at the following approximate altitudes above sea level (Fig. 1):

North of Tunulliarfik:
- At 600 m in the scree slope immediately to the east of the Kvanefjeld plateau, rising to 800 m in the foothills of Ilímmaasaq mountain farther to the east where xenoliths of pulaskite occur in arfvedsonite lujavrite immediately below the contact to the overlying basalts (Rose-Hansen *et al*. 1977).
- At about 550 m on the Kvanefjeld plateau as xenoliths in arfvedsonite lujavrite (Sørensen *et al*. 1969, 1974; Nielsen & Steenfelt 1979; Steenfelt 1981).
- At 400 to 1300 m in Nakkaalaak mountain as a thin sheet between overlying alkali granite and underlying foyaite and sodalite foyaite. It reaches the highest altitude at 1300 m in the north and north-east where the dip of the boundaries between the successive rocks are steep to vertical, and it slopes downwards to 400 to 600 m towards the south and west where the dip is near horizontal (Ferguson 1964; Hamilton 1964; Steenfelt 1981).
- At 700 m in the isolated occurrence to the south of Taseq (Ferguson 1964; Steenfelt 1981).

South of Tunulliarfik, from north to south (Ferguson 1964; Steenfelt 1981; Andersen *et al*. 1988):
- At 230 m as small bodies in foyaite.
- At 530 m as remnants of a layer overlying foyaite.
- At 600 m on the NE side of Nunasarnaasaq in contact with augite syenite and underlain by foyaite, sodalite foyaite and naujaite which at lower levels make up the contact facies of the agpaitic part of the complex.

A straightforward question is, can the variation in altitude be explained by faulting? Only one major fault intersecting the agpaitic rocks, the Lakseelv-Kangerluarsuk fault, has been identified. With the exception of xenoliths in kakortokite and aegirine lujavrite, all occurrences of roof series rocks are located to the north of the Lakseelv fault zone. The gradual transition from pulaskite to naujaite and continuity in outcrops of naujaite make major displacement of the roof series rocks in the area between the fault and Tunulliarfik highly unlikely.

There is a difference in altitude of about 200 m of the location of the contact between naujaite and sodalite foyaite between the two sides of Tunulliarfik. This difference may be explained by sagging of the central part of the complex, a process suggested by Ussing (1912). In any case, there has been little if any fault movement between the north and south sides of Tunulliarfik during or after the formation of the agpaitic rocks (Rose-Hansen & Sørensen 2002).

On the Kvanefjeld plateau, xenoliths of pulaskite, foyaite, sodalite foyaite and naujaite are enclosed in arfvedsonite lujavrite. In one xenolith, the complete transition from pulaskite over foyaite and sodalite foyaite to naujaite takes place over 4-5 m (see the map in Sørensen *et al*. 1974). The pulaskite occurring immediately below the basaltic roof east of Kvanefjeld and in the foothills of Ilimmaasaq mountain is most probably located in its original position. This is a strong indication that also at Kvanefjeld about 1.5 km farther to the west, the uppermost part of the agpaitic rocks originally consisted of pulaskite underlain by foyaite, sodalite foyaite and naujaite which were subsequently intruded by and enclosed in lujavritic rocks. Faulting has displaced the volcanic roof rocks of the Kvanefjeld area downward in the order of 300-400 m prior to the consolidation of the agpaitic rocks (J.G. Larsen 1977). Nielsen & Steenfelt (1979) pointed out that at Kvanefjeld, xenoliths of pulaskite etc. are located at the same altitude as the marginal pegmatite which is found in naujaite. Since pulaskite belongs to a higher stratigraphical level in the complex than naujaite, the xenoliths must accordingly have been displaced downwards. This could explain the difference in altitude of pulaskite between 800 m at Ilimmaasaq mountain and 550 m at Kvanefjeld. However, the transition from pulaskite to naujaite over 4-5 m in one xenolith at Kvanefjeld indicates that pulaskite and naujaite were originally present at nearly the same stratigraphical level. It may therefore be argued that the xenoliths of pulaskite etc. are located more or less in their initial position and that the roof series wedges out towards and across the Kvanefjeld plateau.

Accordingly, north of Tunulliarfik, the roof series rocks from pulaskite to naujaite may originally have formed a continuous sheet which now has the form of an asymmetrical dome. It has its top at about 1300 m in the northern part of Nakkaalaak, a steep north flank and smooth slopes towards the north-west, west, south and south-east. In the north-west part of the sheet which faces Kvanefjeld pulaskite is found at 800 m. The occurrences of pulaskite etc. at Kvanefjeld at 550 m and in the foothill of Ilimmaasaq at 800 m may be considered prolongations of the Nakkaalaak sheet which probably thinned out in that direction. The occurrence of pulaskite etc. located about 700 m to the south of Taseq may be a remnant of the south-western extension of the sheet.

South of Tunulliarfik, at Nunasarnaasaq (Fig. 1), pulaskite is in contact with augite syenite at the altitude of about 600 m and is definitely located in its original position. From that place there is a general downward slope of the pulaskite occurrences toward Tunulliarfik, from 530 m in the southerly to 230 m in the northerly body (Fig. 1, and Steenfelt 1981), i.e. a very steep dip towards the central part of the complex. It should, however, be pointed out that pulaskite in the 230 m occurrence forms xenoliths in foyaite.

In conclusion, the difference in altitude of the roof series rocks across the Ilímaussaq complex cannot be explained by faulting. In the Nakkaalaak-Taseq area, pulaskite, foyaite, etc. are found at the highest altitude of occurrence of nepheline syenites in the Ilímaussaq complex. This and the near vertical boundaries between pulaskite, foyaite, etc. at Nakkaalaak may be explained by steepening of boundaries near the north contact of the complex in accordance with the model proposed by Ussing (1912), but the steep rocks are upwards convex, whereas the layers in Ussing's model are upwards concave. The downward slope of this package of rocks towards the south-west and south-east and the slope towards the north in the part of the complex lying to the south of Tunulliarfik may at least partly be explained by sagging of the central part of the complex. This process cannot, however, account for the downward slope toward Ilimmaasaq and Kvanefjeld to the north-west and west, that is towards the lateral contact of the complex. This suggests that the shape of the roof of the magma chamber was very irregular or that some sort of doming has taken place in this part of the complex, possibly the same process that was the cause of the high-altitude location of the roof series rocks at Nakkaalaak.

The roof series rocks have been eroded to the south of the Lakseelv fault but their former existence is shown by xenoliths of naujaite and foyaite in kakortokite (Bohse *et al.* 1971; Andersen *et al.* 1988). The steeply decreasing altitude of the occurrences of pulaskite, etc. towards the north could indicate that the roof of the magma chamber also in the southern part of the complex had an irregular or updomed form.

The augite syenite

The augite syenite formed an independent intrusion which was successively intruded by alkali granite and alkali syenite in unknown order and by nepheline syenites and was partly replaced by the last-named rocks. Augite syenite masses which are in contact with the country rocks of the complex must be in their original position. This applies especially to the south and west contacts and to the contact againt the roof of volcanic rocks in the apical part of the complex at Nakkaalaaq. Augite syenite enclosed in the agpaitic rocks may either be xenoliths which have fallen in from the roof or walls of the agpaitic magma chamber, or they may be screens of the original syenite intrusion engulfed by the invading rocks and left more or less in the original position.

The kakortokite is rich in augite syenite xenoliths. A fine example is the south contact of the complex at about 400 m a.s.l. in the Laksetværelv (Fig. 1), where small rafts of augite syenite are enclosed in the marginal pegmatite (Fig. 20) and small masses of naujaite are undergoing disaggregation in the matrix of the marginal pegmatite. Large augite syenite screens are enclosed in kakortokite away from the contact. These xenoliths have definitely fallen into the kakortokitic magma from the side walls of the magma chamber, since remnants of the marginal pegmatite are attached to some of them (Bohse *et al.* 1971). Groups of large augite syenite xenoliths in kakortokite and lujavrite occur at Laksefjeld.

Screens and lenses of augite syenite in lujavrite are found near the west contact on the north coast of Tunulliarfik, in the bed of Narsaq Elv and at Kvanefjeld. These xenoliths may have fallen in from the side walls of the magma chamber but could also be parts of the original augite syenite intrusion left *in situ*.

In contrast, augite syenite xenoliths are practically missing in naujaite in spite of the fact that naujaite makes up the largest surface area of the agpaitic rocks of the complex. One reason for this poverty in augite syenite xenoliths may be that xenoliths would sink through the low density volatile-rich naujaitic magma.

The augite syenite masses in the Narsaq Elv bed might be such sunken fragments. However, a large mass of augite syenite in the river bed is in contact with the volcanic country rocks and is considered to be in its original position. It makes up the river bed for a little more than 1 km from the south contact of the complex at an elevation of 100 to 200 m a.s.l. (see Fig. 1 and map in Ferguson 1964). It is intruded by naujaite, aegirine lujavrite and arfvedsonite lujavrite. The contact between this marginal augite syenite and the naujaite immediately to the north is sharp. The augite syenite appears to be intruded and partially enclosed by naujaite (Fig. 27), a further observation in support of the interpretation that this augite syenite rests more or less *in situ*. Northwards in the river bed there are some small occurrences of augite syenite, which are associated with lujavrites. Most of these are shown on Ferguson's map, but the northernmost occurrence on his map has not been verified.

In the south-western part of the Kvanefjeld plateau (Fig. 1), large screens of augite syenite occur at altitudes from about 200 to 500 m a.s.l. on the slope towards the Narsaq Elv valley. They are enclosed in and veined by arfvedsonite lujavrite containing xenoliths of augite syenite, alkali syenite, naujaite and volcanic rocks. The augite syenite screens are underlain by naujaite in the wall facing the valley, and the overlying lujavrite encloses naujaite xenoliths. Twenty-two of the seventy bore holes on the

Fig. 27. Contact between augite syenite (brownish, foreground and left of hammer) and naujaite (right side and top). Narsaq Elv river bed about 100 m a.s.l.

Kvanefjeld plateau and immediately to the east of the plateau have augite syenite xenoliths in lujavrite and syenite dykes intersecting the roof basalts. They are confined to the south-western part of the plateau where augite syenite is exposed at the surface. Therefore, the northern contact of the former augite syenite massif probably coincides more or less with the limit of augite syenite xenoliths in the south-western part of the plateau. The augite syenite xenoliths occur in the depth interval from 20 to 400 m in the drill cores, i.e. from about 500 to 100 m a.s.l., in altitude overlapping the occurrences in the river bed.

On the Kvanefjeld plateau, the identical orientation of structures in neighbouring augite syenite xenoliths suggests that the augite syenite originally constituted one or a few large bodies (Nielsen & Steenfelt 1979) which now rest more or less *in situ*. In this case the overlying agpaitic rocks have transgressed the original contact of the augite syenite body. Thirty of the Kvanefjeld drill cores have naujaite xenoliths in the depth interval of 25 to 450 m in the core. Seventeen cores have augite syenite as well as naujaite; naujaite occurs at a higher level than syenite in seven, at a lower level in four and more or less at the same level in six cores. It is difficult to decide whether these augite syenite and naujaite xenoliths were formed by falling in from the walls of the lujavritic magma chamber or they are screens of the original augite syenite intrusion and naujaite left more or less *in situ*.

There is a difference in altitude of about 500 m between the augite syenite occurrences at Kvanefjeld and the part of the Nakkaalaak augite syenite sheet facing Kvanefjeld. The difference may indicate faulting prior to the emplacement of the agpaitic rocks or foundering of augite syenite screens in the agpaitic magma chamber, but the Nakkaalaak augite syenite sheet is conformable with the Nakkaalaak-Taseq roof series package and with the shape of the roof of the agpaitic magma chamber. This being the case, the sloping Nakkaalaak augite syenite sheet may originally have been continuous with the Kvanefjeld augite syenite. Thus, it is possible that the augite syenite xenoliths of the Kvanefjeld plateau may be remnants, resting more or less *in situ*, of the augite syenite intrusion which is inferred to have occupied a large volume prior to the emplacement of stage 3 of the complex. A look at the geological maps (Fig. 1; Ferguson 1964; Andersen *et al*. 1988) shows a bewildering scatter of augite syenite remnants but nevertheless a tendency to separate into three major groups: a southern, a western and a roof group. It is difficult to determine the primary outlines of the augite syenite intrusion and the number of intrusions.

Alkali granite, alkali syenite and microsyenites

The alkali granite was emplaced after the augite syenite and before the pulaskite. Two sheets of granite are preserved below the outliers of the volcanic roof of the complex. The quartz syenite is most probably formed by reactions between the granite and the nepheline syenitic magma.

The alkali syenite definitely intrudes augite syenite and is intruded by arfvedsonite lujavrite. Its age relation to the alkali granite and to kakortokite and the roof series rocks is unknown due to the lack of contacts. However, microsyenites, which are mineralogically and chemically related to the alkali syenite, form sheets intruding kakortokite, aegirine lujavrite and naujaite and are intruded by arfvedsonite lujavrite. The microsyenites are considered to be of external origin. This may also be the case for the alkali syenite (Rose-Hansen & Sørensen 2001).

The nepheline syenites

It is still under debate whether stage 3 in the evolution of the Ilímaussaq complex, the nepheline syenites, involved one, two or perhaps more magma pulses. It is generally agreed that the roof series rocks were formed in the upper part of the complex by crystallization of the first (and perhaps only) magma pulse. They display sharp contacts against their country rocks and have very few xenoliths of the earlier rocks. Naujaite apophyses intrude the country rocks at Kvanefjeld and Nakkaalaak.

The view that the roof series crystallized downward in the order pulaskite, foyaite, sodalite foyaite and naujaite is based on unequivocal observations. Pulaskite is in contact with augite syenite at Nunasarnaasaq and with alkali granite and quartz syenite at Nakkaalaaq. It contains angular augite syenite xenoliths (Fig. 22) close to the contact NE of Nunasarnaasaq. The interpretation that pulaskite was the first rock to crystallize faces, however, the problem that it only occurs in the uppermost part of the complex and not along the sidewalls where one would expect an intruding hot magma to quickly cool and crystallize. The possibility, that pulaskite originally formed along also the side walls, and that this pulaskite rim has subsequently been eroded and disaggregated in the volatile-rich magma, cannot be excluded, but pulaskite xenoliths have not been observed in the underlying contact rocks. The pulaskite xenoliths in foyaite described by Steenfelt (1981) occur in the roof zone of the complex. The missing side-wall pulaskitic contact facies may be related to the flat disc-shape of the magma chamber. A thin shell of pulaskite was rapidly formed along the upper contacts where heat loss was strong through the roof of the magma chamber. Heat loss was weaker through the short marginal walls. This combined with accumulation of volatile-rich residual liquids beneath the crystallization front and the impermeable lid of the volcanic roof rocks delayed or arrested crystallization of the uppermost part of the magma. The increasingly volatile-rich magma cooled slowly and crystallized successively as the ambient temperature fell below the liquidus temperature at the existing conditions. This resulted in the formation of horizons of foyaite, sodalite foyaite and naujaite across the entire complex, from side contact to side contact (Larsen & Sørensen 1987). The volatile-rich melts were in contact with the adjacent country rocks for extended periods causing intense fenitisation of these.

South of Kangerluarsuk, the naujaite masses in the west contact of the complex between augite syenite and marginal pegmatite show that naujaite was formed some hundred metres below the present main zone of naujaite and that it is older than kakortokite. This and the rafts in lujavrite and kakortokite indicate that, originally, the naujaite horizon was considerably thicker than the now exposed 600 m.

The lower surface of the main mass of naujaite, i.e. the ceiling of the lujavritic magma chamber, is highly irregular. Thus, at the head of Kangerluarsuk, naujaite is exposed in the coast. It is in contact with aegirine lujavrite without any intervening arfvedsonite lujavrite; the latter is found farther to the east at a higher stratigraphical level in the the complex and mainly in the form of intrusive sheets in the naujaite roof (Fig. 1 and map of Andersen *et al.* 1988; Rose-Hansen & Sørensen 2002). These features indicate that at the head of Kangerluarsuk there is a lateral wedging-out of the lujavrite sequence and a depression in the boundary between naujaite and lujavrite. The cross-section of the complex provided by the north coast of Tunulliarfik (Fig. 23) also shows domes? and 'depressions' in the boundary between lujavrite and naujaite. With the exception of the marginal contact zones, where the lujavrite lamination and layering are steep, the igneous lamination and layering of the main mass of kakortokite-lujavrite are almost horizontal and are accentuated by the conformably enclosed naujaite rafts. The layering in the enclosing lujavrite drapes over the naujaite rafts (Fig. 25, Bailey *et al.* 2006).

The one-magma batch model presents a simple interpretation of the formation of stage 3 of the Ilímaussaq complex: lujavrites formed from the residual magma which was sandwiched between the downward growing roof series and the upward growing kakortokites (Ferguson 1964; Gerasimovsky 1969; Engell 1973; Larsen & Sørensen 1987).

The lowermost part of the naujaite horizon and naujaite xenoliths in kakortokite are strongly altered which indicate that the naujaite had been exposed to prolonged contact with the magma. This may be related to a temporary suspension of crystallization caused by the build-up of Na, volatiles and residual elements below the naujaite roof which lowered the liquidus temperature below the ambient temperature of the magma (Steenfelt & Bohse 1975). New gradients of temperature, composition and density were established in the magma chamber and crystallization shifted to the stagnant layer on the floor of the chamber resulting in the formation of the layered kakortokites which graded upwards into aegirine lujavrite (Larsen & Sørensen 1987).

That there was a break between the crystallization of naujaite and kakortokite is emphasised by the existence of the marginal pegmatite which represents the initial magma of the kakortokite-lujavrite sequence. The marginal pegmatite zone has sharp contacts against the country rocks and encloses augite syenite and naujaite xenoliths. It passes gradually into the main zone of layered kakortokite and lines the marginal contact of the overlying aegirine lujavrite I and II. This combined with the fact that the kakortokite layers span the complex from contact to contact and that there is a gradual transition into aegirine lujavrite suggest that these rocks formed by crystallization of one batch of magma.

The main mass of lujavrite is sandwiched between naujaite and kakortokite, but lujavrites also transgress the roof series rocks and the marginal contacts of the complex, including the augite syenite rim, and some occurrences may have been derived from separate pools of lujavrite magma (Sørensen *et al.* 2006)

The kakortokites are beyond any doubt cumulates and so are the aegirine lujavrites and the main mass of the overlying arfvedsonite lujavrites. Internal intrusive relationships are rare in the main kakortokite-lujavrite sequence. The density of the naujaite xenoliths is almost identical to that of the Fe-rich lujavritic magma which led Ussing (1912) to conclude that the naujaite rafts could not sink in the magma and that they therefore rest more or less in their original position. Planar structures in adjacent naujaite rafts are almost parallel and also parallel to the lamination and layering of the enclosing lujavrites. This, the cumulitic lujavrite textures, the near horizontal layering and lamination, and the scarcity of internal intrusive relations, lead to the following model for the emplacement of the lujavrite body: the naujaitic roof of the lujavritic magma chamber was fractured and infiltrated by the rising magma. The main mass of lujavrites was emplaced by slow upward layer for layer growth making room for the rise by piecemeal stoping of the overlying naujaite which was successively enclosed as rafts in the lujavrite (Sørensen et al. 2006). In the waning stage of intrusion, naujaite slices were detached from the roof but were not enclosed as rafts in the lujavrite because further rise of the magma was arrested or the melt had become so viscous that the rafts could not settle in the consolidating lujavrite. Therefore, the uppermost naujaite xenoliths show a somewhat irregular arrangement (Fig. 23).

Discussion

The survey of the published views on the formation of the Ilímaussaq complex shows that there are difficulties explaining why pulaskite only occurs in the uppermost part of the complex, how the huge amount of sodalite in naujaite could form, the origin of the kakortokite-lujavrite sequence and the topography of the roof of the agpaitic magma chamber. These questions will be discussed in the light of the observations and data reported in the present paper.

Most of the rocks of the Ilímaussaq complex are cumulates and chill zones are very rare. It is in this context of interest to examine the information on melt compositions and evolution provided by the homogeneous massive-textured rocks from the marginal pegmatite and the other homogeneous massive rocks described in the present paper.

The significance of the marginal pegmatite in the Ilímaussaq complex

The marginal pegmatite is the contact facies of the kakortokitic lower part of the magma chamber and locally of the naujaitic part near the roof of the complex. It consists of massive rocks veined by short pegmatites. The pegmatites were most probably formed in pockets of volatile-rich melt in the almost completely consolidated contact facies, but the locally developed dyke-like pegmatites may have been emplaced in cooling cracks by melts released from the consolidating kakortokite and lujavrite.

At the west contact of the complex, the north side of the Lakseelv-Kangerluarsuk fault zone is down-faulted at least 500 m relative to the south side. This explains that on the south coast of Kangerluarsuk, the marginal pegmatite passes gradually into the layered kakortokite, whereas on the north coast of this fjord, the marginal pegmatite is in contact with aegirine lujavrite I. The massive-textured sample 104363A collected at about 200 m elevation south of the fault represents an at least 300 m deeper level in the complex than sample 104361A collected on the north shore of Kangerluarsuk.

Tunulliarfik was the seat of major faulting before, but not during or after the emplacement of the agpaitic rocks. In accordance with this interpretation, the aegirine lujavrites occurring in contact with marginal pegmatite at sea level on the south and north coasts of Tunulliarfik have been identified as aegirine lujavrite II and thus most probably belong to one and the same stratigraphical level of the complex. The marginal pegmatite 104361A from the north coast of Kangerluarsuk is in contact with aegirine lujavrite I, i.e. a lower stratigraphical level than 109303 from the north coast of Tunulliarfik. Sample 104363A represents the still deeper kakortokite level and 104380A the naujaite level of the complex.

In summary, the matrix of the marginal pegmatite was the first agpaitic rock to crystallize in the lower parts of the complex made up of kakortokite and aegirine lujavrite. At higher levels in the complex, naujaite, sodalite foyaite and arfvedsonite lujavrite are in direct contact with augite syenite and the volcanic country rocks. Pulaskite is in contact with the augite syenite rim at Nunasarnaasaq and with sheets of alkali granite and quartz syenite at Nakkaalaak (Fig. 1). A zone of marginal pegmatite occurs between naujaite and the volcanic rocks in the northern part of Kvanefjeld and north of Nakkaalaak. The matrix of the Kvanefjeld occurrence is made up of naujaite and of a homogeneous massive-textured agpaitic rock.

Information from the chemical analyses of the matrix of the marginal pegmatite

Chemical analyses of the massive-textured matrix of the marginal pegmatite are compiled in Table 2 which also presents analyses of tephriphonolite and phonolite dykes (153099 and 153200) intersecting the country rocks of the complex (quoted from Larsen 1979).

Ussing (1912) considered sodalite foyaite and the average composition of the agpaitic rocks as measures of the composition of the initial agpaitic melt. The average compositions of the agpaitic Ilímaussaq rocks listed in Table 4 are quoted from Ussing (1912), Gerasimovsky (1969) and Ferguson (1970a, b). Table 4 also presents analyses of interstitial glass in nepheline syenite xenoliths in phonolitic tuff at Tenerife (Wolff & Toney 1993) and of an agpaitic dyke (42475) intersecting the country rocks south of the Ilímaussaq complex. The interstitial glass and the dyke proxy the composition of agpaitic melts.

The analyses of the matrix rocks 104361A, B and 109303 differ from the average composition of the Ilímaussaq nepheline syenites of Table 4 in having higher SiO_2, Fe_2O_3, K_2O and F and lower TiO_2, FeO, Na_2O and Cl contents, whereas 104380A is rather close to this composition. The chemical analyses of 104363A and the interstitial glass from Tenerife show some similarities. Higher contents of volatile components like Na_2O, Cl and F in the glass may be ascribed to the much more rapid quenching of the glass than of the marginal pegmatite rock.

The matrix rocks 104361A, B, 109303 and 104380A differ from most of the Ilímaussaq rocks of Table 5 first of all in their relatively elevated contents of Cs, Ba and Sr. The Cs content can be at least partly explained by analcimisation of the massive-textured matrix rocks (Bailey *et al.* 2001). 104361A shows close resemblance to kakortokite with respect to major elements, but it has higher contents of Ba and Sr and lower contents of most other trace elements and is thus less evolved than kakortokite. When compared with the kakortokite, 109303 is enriched in the analcime and microcline components Na_2O, Al_2O_3, H_2O, Cs and K_2O and Rb and it has lower contents of most trace elements. Ba is lower than in 104361A and kakortokite and at the same level as in aegirine lujavrite.

The massive-textured matrix rocks 104363A and 104380A are practically unaltered, whereas in 104361A, B and 109303, nepheline is altered to analcime and sericite, the amphibole to aegirine and pigmentary material, and eudialyte to zircon etc., whereas K feldspar appears to be largely unaltered. It is a general feature that coarser-grained Ilímaussaq nepheline syenites – which are characterised by an open foyaitic texture and by a long temperature interval between liquidus and solidus – have been exposed to alteration of the type described here. How will this influence the chemical composition of the rocks?

The major element compositions (Table 2) and element ratios such as Na/K and Na/Ca of 104361A (1.2 and 2.7) are very close to those of dyke 153099 (1.2 and 3.0) (Table 6). The dyke proves the existence of melts of this composition. If the matrix rocks originated from such melts, alteration has not changed the bulk composition of the rocks and appears to have taken place in a closed system. The level of elements such as Ba, Sr, Zr, REE and Nb is, however, higher in 104361A than in 153099. The chemical composition of dyke 153200 (Table 2) represents an intermediate stage between these rocks.

The CIPW weight norms of the altered

rocks, 104361A and 109303, differ from those of the unaltered samples 104363A and 104380A (Table 3) in the lack of *ns* and the presence of *hm*. 104363A has higher Na/K, Na/Ca and Na/Zr ratios (Table 6) than 104361A. This suggests a loss of Na in 104361A and alteration in an open system.

Elements such as Zr, REE, Y, Nb, U and Th are at the same level in 104363A and 104361A and B, but the Pb and Zn contents are higher and contents of Rb, Ba and Sr lower in 104363A. 104361A and 104363A have a number of identical or almost identical element ratios: K/Rb, K/Ba, Ca/Ba, Ca/Sr, Ba/Sr, Ba/Rb, La/Y, Th/U, Zr/U, Zr/La, Zr/Y and Zr/Nb, whereas the Na/K, Ca/Zr, Zr/Th, Zr/Ba and Zr/Sr ratios are different (Table 6). It appears that Zr covaries with U, La (REE), Y and Nb and not with Ca, Ba, Sr and Th and that there is a covariation of K, Ca, Rb, Ba and Sr and also of REE and Y. This may indicate a separation of Ca, Ba and Sr from Zr caused by the alteration of the eudialyte crystals of 104361A, B, whereas there was no separation of REE, Y, Nb and U from Zr which is in agreement with results of earlier investigations (Andersen *et al.* 1981b; Bailey *et al.* 2001, Rose-Hansen & Sørensen 2002).

The similarity in chemical composition between matrix rocks 104361A and 104363A and the fact that 104363A is practically unaltered makes 104363A a likely representative of the composition of its parental melt. The matrix rock 109303 from the north coast of Tunulliarfik is less altered than 104361A and B. It deviates from these rocks and from 104363A in having higher contents of MgO, Na_2O, Al_2O_3, H_2O and most trace elements and, with the exception of K/Rb, Zr/U, Zr/La, Zr/Y and Zr/Nb, different element ratios.

In matrix rock 104380A from Kvanefjeld, the contents of elements such as Zr, REE, Nb are slightly higher than in 104361A, B, 104363A and 109303 from lower stratigraphical levels in the complex.With regard to major elements, 104380A is very close to sodalite foyaite but it has higher contents of Zr, CaO, Fe and most trace elements and lower contents of alkalis and Cl than this rock which reflects the high eudialyte content in 104380A and high sodalite content in sodalite foyaite (Table 5). The low Cl content in 104380A is unexpected at this level of the complex. Sodalite has not been observed in this rock. If originally present, sodalite must have been substituted by analcime. The element ratios of 104380A are closer to those of matrix rocks 104361A, B, 104363A and the kakortokite-lujavrite sequence than to the associated sodalite foyaite and naujaite (Table 6). 104380A may represent the composition of the parental magma at this level of the complex.

The differences between the compositions of 104361A and 104363A and between these rocks and 109303 and 104380A may be explained by their different stratigraphical levels, mentioned from the top downwards: 104380A, 109303, 104361A and 104363A. Different degrees of alteration and local heterogeneity may also have played a role. Thus, the 109303 nepheline syenite is very heterogeneous. Its nepheline content varies from 20% to more than 50%, the analysed sample contains about 20% nepheline.

According to Ferguson (1970a, b), the K/Rb ratios of average analyses of the various Ilímaussaq agpaitic rocks vary within narrow limits and decrease with the inferred order of formation of the rocks. Bailey *et al.* (2001) found correlation between fractionation trends and chronology, thus the K/Rb ratio decreases from 460 to 35 during evolution of the complex (129 to 53 in Table 6). Andersen *et al.* (1981b) established a geochemical stratigraphy for the kakortokite-lujavrite sequence based on the Zr/U and Zr/Y ratios. From kakortokite to late lujavrites, Zr/U decreases from

1200 to 9.1 (581 to 42.5 in Table 6) and Zr/Y from 18.2 to 2.8 (19.5 to 8.2 in Table 6). Bailey *et al.* (2001) distinguished discrete Zr/U arrays for the roof series rocks, pulaskite, foyaite, sodalite foyaite and the upper part of naujaite falling along one array, the lower part of the naujaite plots along another array which has three times lower U for a given Zr content. The kakortokite-lujavrite sequence shows different Zr/U arrays. This was interpreted to be a result of stratification of the agpaitic magma. Bailey *et al.* (2001) pointed out that a fall in the Ba/Rb ratio is reversed from naujaite to kakortokite and that median Sr contents decrease from augite syenite through the roof series but that there is a significant increase from naujaite to kakortokite. Such a reversal in trend from naujaite to kakortokite was also demonstrated by Ferguson (1970b) for the Ba/K, Ba/Rb, Ba/Sr and Sr/Ca ratios.

The major element contents of the black rock from the mini-scale model pegmatite (107724) are rather close to the composition of kakortokite (Table 5) but, with the exception of Rb and Ba, with much lower contents of trace elements. It is rich in microcline and fluorite and without eudialyte which explains aberrant element ratios. The location of this rock in the central and lower parts of the mini-scale model pegmatite in kakortokite recalls the position of the lujavrites in the whole complex. The black rock differs, however, from the lujavrites in having very low contents of most trace elements contrary to the enrichment that is characteristic for the lujavrites (Table 5). This may be explained by the copious crystallization of eudialyte in the middle and lower parts of the pegmatite which deprived the pegmatitic melt of Zr, Nb, REE, etc. Quenching of the residual melt would accordingly result in a rock devoid of eudialyte and other rare-element-containing minerals. Ca combined with F to form fluorite.

In the one batch/sandwich model, the agpaitic rocks of the complex are thought to have been formed by consolidation of one batch of magma. Pulaskite and foyaite, the earliest rocks of stage 3, have low Ba and Sr contents but higher than the succeeding sodalite foyaite and naujaite which have extremely low contents. These very low contents suggest that Ba and Sr were preferentially taken up by crystallizing minerals or were accumulated in the magma during the formation of the roof series. Ba is partitioned into feldspar rather than into eudialyte, but the naujaite feldspar has very low Ba contents (Bailey personal information). Eudialyte may not be the main carrier of Sr (see p. 31). The low Ba and Sr contents in the roof series rocks may perhaps therefore be attributed to the retention of Ba and Sr in the magma during formation of the roof series because no minerals were fractionating Ba and Sr from the melt. In any case, kakortokite and lujavrites have higher Ba and Sr contents than the roof series rocks and so have the matrix rocks of the marginal pegmatite. The higher Ba and Sr contents in these rocks might be caused by a change in the partition coefficients of Ba and Sr at the onset of formation of kakortokite. Larsen (1979) studied the element distribution between crystals and melt in trachytic and phonolitic dykes adjacent to the Ilímaussaq complex and in sodalite foyaite from Ilímaussaq. The partition coefficients decrease with increasing peralkalinity for REE, Y, Zr, Nb, Hf, Pb and Th and increase for Ba and Cs, whereas the Rb and Sr values are constant. This could perhaps partly explain the observed distribution of Ba and Sr in the roof series rocks.

The distribution of Ba and Sr is one of a number of examples of mineralogical and geochemical discontinuities between naujaite and kakortokite-lujavrite.

Other examples are: (1) The very low

whole-rock U contents decrease from sodalite foyaite to naujaite which is paralleled by decreasing U contents in the eudialyte of these rocks. Complexing of U ions may have retained U in the magma (Bohse *et al.* 1974; Steenfelt & Bohse 1975). The U contents of the kakortokite-lujavrite sequence show the opposite trend and increase upwards through the sequence in rocks and in eudialyte. The eudialyte of the roof series is generally interstitial whereas it in kakortokite-lujavrite is a cumulus phase, i.e. it crystallized late in the roof series rocks, early in kakortokites-lujavrites. This may explain the different uptake of U in eudialyte in the two rock sequences. The decreasing contents of U, Ba and Sr from pulaskite to naujaite could also be referred to exhaustion of very low initial contents of these elements during crystallization of the roof series. The increase of U upwards in the kakortokite-lujavrite sequence, in which eudialyte is a liquidus phase, may be explained by higher initial U contents of the magma and by U entering eudialyte only in limited amount. Therefore, U continued to concentrate in the crystallizing kakortokitic melt (Andersen *et al.* 1981b).

(2) Engell (1973) showed that in the Na_2O-Al_2O_3-Fe_2O_3-SiO_2 system kakortokite and lujavrite diverge from the trend of the roof series rocks and he expressed doubts concerning the view that the agpaitic rocks were formed by crystallization of one batch of magma in a closed system.

(3) Similarly, in the system SiO_2-Al_2O_3-(Na_2O + K_2O) the kakortokite-lujavrite trend deviates from the augite syenite-foyaite-sodalite foyaite-naujaite trend (Macdonald 1974).

(4) There are differences in ε_{Nd} values between kakortokite-lujavrite on the one side and the roof series rocks on the other (Marks *et al.* 2004). Stevenson *et al.* (1997) reported uniform ε_{Nd} values in the roof series rocks and deviating values in kakortokite and lujavrite.

(5) The element ratios presented in Table 6 show some scatter, but ratios such as Na/K, K/Rb, La/Y, Zr/Hf, Zr/La, Zr/Y, Zr/Nb, Nb/Ta and Th/U display remarkably little variation throughout the Table. However, closer scrutiny of the ratios suggests a divison of the rocks of stage 3 into two separate trends of evolution: (a) From pulaskite over foyaite to sodalite foyaite (Na/Ca, K/Ba, Ca/Sr, Rb/Sr, Zr/Y, Ca/Zr, Nb/Ta, Zr/Ba, Zr/Sr). There appears to be a break between sodalite foyaite and naujaite which may be ascribed to the marked change in the crystallization conditions of the roof series when naujaite formed by flotation of sodalite crystals. (b) From marginal pegmatite over kakortokite and aegirine lujavrite to arfvedsonite lujavrite (Na/K, Ca/Ba, Ca/Sr, Rb/Sr, Zr/Sr, K/Rb, Na/Ca, K/Ba, Nb/Ta, Th/U, La/Y, Zr/U, Zr/Y, Zr/Ba).

These discontinuities invite speculation that naujaite and kakortokite were formed from successive magma pulses as proposed by Larsen (1976) and Bailey *et al.* (1981) and corroborated by the distinction of separate series of rocks by means of their Zr/U ratios (Bailey *et al.* 2001). The Zr/U ratios in Table 6 do not contradict this proposal but are too few to examine its validity. The kakortokite-lujavrite stage then represents a new magma injection which was characterised by higher contents of Ba and Sr and of Zr, REE, U, Nb, etc. than the initial melt of the roof series.

In summary: The marginal pegmatite was the first agpaitic rock to form at the stratigraphical level of kakortokite and aegirine lujavrite. At the lowest exposed level, it passes laterally into kakortokite; at higher levels it is intruded by aegirine lujavrite I and II. Its matrix is more evolved than pulaskite and related to but less evolved than kakortokite. Marginal pegmatite is missing along most of the contacts above the aegirine lujavrite but is, however, found near the roof of the complex where its matrix is chemically related to but more evolved than sodalite foyaite and may represent the composition of the parental magma at this level of the complex. Regarding the initial magma of the kakortokite-lujavrite sequence, it may be concluded that it either had a major element composition comparable to that of 104361A and dyke 153099, i.e. with comparatively low Na and K, or it was richer in Na as exemplified by 104363A. It should be emphasised that in both scenarios the initial magma of the kakortokite-lujavrite suite had remarkably high contents of Ba, Sr, Zr, REE, etc.

Information from other Gardar complexes

Augite syenite is an important constituent of the igneous complexes of the Gardar province (Sørensen 1966, 2003; Emeleus & Upton 1976; Upton & Emeleus 1987; Upton *et al.* 2003). In most of the complexes a weakly oversaturated syenite is associated with silica-oversaturated syenites and granite. However, two complexes provide information about the association of silica-undersaturated augite syenite and nepheline syenite.

In the Motzfeldt centre, one of the four major units of the Igaliko complex in the easternmost part of the province (Emeleus & Harry 1970), there is a gradual evolution from a marginal augite syenitic facies through foyaite to eudialyte-nepheline syenite. These rocks are intruded by lujavrites which mineralogically and chemically recall the Ilímaussaq lujavrites and have 9.40-10.92% Na_2O, 40-263 ppm Ba, 26-73 ppm Sr and 1040-4230 ppm Zr (Jones & Larsen 1985).

The second occurrence is the Tugtutôq Older Giant Dyke west of the Ilímaussaq complex. This about 700 m wide dyke is composite with a marginal zone of olivine gabbro, an intermediate zone of syenogabbro and a core of syenitic rocks (Upton *et al.* 1985, 2003). The syenitic core shows from west to east an apparently continuous evolution from augite syenite to foyaite, the former contains 2105 ppm Ba and 255 ppm Sr, the latter 35 ppm Ba and 21 ppm Sr. The foyaite is regarded as the uppermost exposed part of the core of the dyke. The giant dyke shows a spatial and genetic association of gabbro and syenite. Gabbro was emplaced first and was followed later by the syenitic core, most probably from the same deep-seated fractionating magma chamber.

In both complexes there is a gradual transition from augite syenite to nepheline syenite, whereas in the Ilímaussaq complex there is a time gap between the emplacement of the augite syenite and the nepheline syenites.

The existence of a compositionally layered magma chamber is demonstrated in the Kûngnât complex in the north-western part of the Gardar province (Fig. 1) (Stephenson & Upton 1982; Upton *et al.* 2003) where expulsion of melts from the top of the chamber downwards produced the succession quartz syenite, basic syenite, syenodiorite and gabbro.

Synthesis and conclusions

(1) An augite syenite massif was intruded successively by alkali syenite, alkali granite and by nepheline syenites resulting in the formation of the composite Ilímaussaq alkaline complex. Augite syenitic and nepheline syenitic melts are thought to form in a fractionating alkali basaltic magma chamber in the deep crust as outlined by Larsen & Sørensen (1987) and further developed by Stevenson *et al.* (1997), Markl *et al.* (2001) and Marks *et al.* (2004). The nepheline syenitic melts most probably originated in a magma chamber that had acquired an augite syenitic composition (Sørensen 2003). This interpretation is corroborated by the association of augite syenite and nepheline syenite in two other Gardar intrusive complexes, the Tugtutôq Older Giant Dyke complex (Upton *et al.* 1985) and the Motzfeldt centre of the Igaliko complex (Jones & Larsen 1985).

The coexistence of silica-undersaturated and -oversaturated rocks is a rare phenomenon. One example is the Agua de Pau volcano, São Miguel, Azores Islands where the coexistence is referred to the parental basaltic magma straddling the critical plane of undersaturation in the basalt tetrahedron (Ridolfi *et al.* 2003). Small changes in the physicochemical parameters in the magma chamber decided whether *ne* normative or *q* normative rocks were formed. A second example is the spatial coexistence of granite, syenite and nepheline syenite in the Mykle region in the southern part of the Oslo igneous province, South Norway (Andersen & Sørensen 1993). The nepheline syenite was emplaced in a large larvikite massif and was subsequently intruded by first syenite and thereafter by granite. It is now found only as xenoliths in syenite and granite.

In the case of the Ilímaussaq complex, assimilation of country rocks appears to be the most probable explanation of the coexistence of nepheline syenites and granite. On the basis of Nd isotope studies, Stevenson *et al.* (1997) proposed that the granite was formed by assimilation of country rocks in the augite syenitic magma. This conclusion is strengthened by studies of the Nd- and O-isotope systems by Marks *et al.* (2004) who stated that parts of the augite syenitic magma became contaminated with lower crustal material which resulted in the formation of granite. This explains the geochemical similarities between the granite, augite syenite and agpaitic rocks.

(2) The nepheline syenites of the complex are divided into a roof series, a lujavritic intermediate series and a floor series.

The roof series crystallized downward in the order: pulaskite, foyaite, sodalite foyaite and naujaite, of which the two last-named are agpaitic. The rocks grade into each other and thus appear to be products of one largely continuous crystallization process which took place under an impermeable roof that prevented volatiles from escaping.

The naujaitic part of the roof series is now more than 600 m thick. It was originally thicker because the arfvedsonite lujavrite and aegirine lujavrite enclose rafts of naujaite. Bore holes II and VI terminate in large naujaite masses, more than 80 m thick, beneath more than 120 m lujavrite. Additionally, naujaite occurs along the south contact of the complex between the marginal pegmatite and the augite syenite rim some hundred metres below the present lower boundary of the roof series. The magma chamber most probably had its maximum vertical extent at the naujaite stage.

(3) Pulaskite, the first rock to crystallize in the roof series, is cumulitic.The pulaskite magma reacted with the alkali granite resulting in elevated SiO_2 contents. Therefore, its chemical composition is not entirely equivalent to the composition of the parental melt of the roof series.

At Kvanefjeld, i.e. near the roof of the complex, the matrix of the marginal pegmatite consists of naujaite and a massive-textured rock. With regard to major elements, the last-named rock corresponds to sodalite foyaite and may represent the composition of the initial magma at this level of the complex. The contents of Ba and Sr and of most trace elements are higher than in the roof series rocks and similar to the levels found in the matrix of the marginal pegmatite surrounding kakortokite and aegirine lujavrite I and II.

(4) The marginal contact facies of the roof series shows a vertical zonation. Pulaskite is in contact with augite syenite and the country rocks only in the top part of the complex. Downward, successively foyaite, sodalite foyaite and naujaite constitute the contact facies, in places transgressed by lujavritic intrusions.

A stratified magma chamber was invoked by Larsen & Sørensen (1987) and Sørensen & Larsen (1987) in explaining the formation of the layered kakortokite sequence, by Bailey (1995) in explaining layering in aegirine lujavrite and by Bailey *et al.* (2001) in interpreting the Zr/U and Zr/Y stratigraphy of the kakortokite-lujavrite sequence. Layering of the roof series magma was, however, rejected by Larsen & Sørensen (1987: 479). In this magma body, the volatile content would be highest in the uppermost part and one should expect crystallization to take place from the floor upwards. But the roof series was definitely formed downward from successively more differentiated, volatile-rich and less-dense melts which would not have been stable on top of each other. Therefore, the roof series rocks most probably were formed by slow consolidation of a well-mixed magma, in conformity with the observation that the pulaskite, foyaite, sodalite foyaite and the upper part of the naujaite fall along one Zr/U array (Bailey *et al.* 2001). An additional argument against a stratified magma chamber during the formation of the roof series is provided by sodalite contents of up to 75 vol.% in large masses of naujaite which would demand accumulation of sodalite formed over a considerable depth interval, possibly the entire magma chamber. Fluid inclusion studies indicate that naujaitic sodalite began crystallization already when the magma was on its way up from the crust (Markl *et al.* 2001). In that case, the system could not be closed downward or naujaite may have been formed from an independent magma pulse.

Against a separate origin of the naujaite speaks that there is a clear mineralogical transition from pulaskite over foyaite and sodalite foyaite to naujaite which is seen in decreasing contents of the primary mafic minerals and the simultaneous development of the agpaitic mineralogy (Larsen 1976, 1977; Markl *et al.* 2001). This suggests that the roof series most probably formed by slow consolidation of a magma that was continuously modified by ascending volatile-rich residual melts and by fractionation of minerals. Pegmatite horizons in foyaite and naujaite mark interruption of the crystallization of these rocks (Sørensen & Larsen 1987: 479). The same probably applies to alternating sheets of naujaite and sodalite foyaite in the top of the naujaite, to pulaskite xenoliths in foyaite and to naujaite intersecting pulaskite (Steenfelt 1981: 48).

(5) The exposed part of the floor series, the kakortokites, occurs as cumulates on the floor of the magma chamber. The marginal pegmatite shows a gradual transition into kakortokite and constitutes its contact facies which intruded not only the augite syenite but also the naujaite that occupied a part of the magma chamber prior to the formation of the kakortokite. The chemical composition of the homogeneous, massive-textured matrix of the marginal pegmatite is thought to be equivalent to the 'kakortokitic' primary melt. As is the case for other Ilímaussaq massive-textured rocks, which crystallized under retention of the interstitial residual melts, late- and post-magmatic alteration may have changed the chemical composition. The chemical analysis of the unaltered sample 104363A, which is part of the matrix or formed by quenching of a pegmatitic part of the marginal pegmatite, is, however, so similar to the analyses of the altered matrix samples that it is felt justified to conclude that alteration had moderate influence on the chemical composition of the rocks and that, apart from the possible loss of first of all Na, the matrix of the marginal pegmatite represents the composition of an early stage of evolution of the initial melt of the kakortokite-lujavrite sequence. The melts, from which these rocks formed, are characterised by elevated contents of Na, Fe, Ba, Sr, REE, Zr, Hf, Nb and Ta. The differences in composition of samples 104363A, 104361A, 109303 and 104380A may be related to variation in composition with height in the complex. This composition differs from that of the initial pulaskitic melt of the roof series in being agpaitic and in crystallizing a different set of minerals. Similar to the naujaite, the last-formed rock of the roof series, the initial melt at the kakortokite-lujavrite stage crystallized K-feldspar, nepheline, arfvedsonite, aegirine and eudialyte. Sodalite and aenigmatite, which are important minerals in the naujaite, have not been identified in the matrix of the marginal pegmatite from low stratigraphical levels. It cannot be excluded that these minerals have been present but they have in that case not survived the alteration of the rocks. Sodalite and aenigmatite occur in the lower part of the kakortokites and aenigmatite in the fine-grained black rock in the mini-scale model pegmatite. The matrix rock from Kvanefjeld, 104380A, which may represent the composition of the naujaitic parental melt or a precursor to this melt, contains aenigmatite, but not sodalite.

(6) When looking at the geological map and at cross sections of the complex, the immediate impression is that the agpaitic rocks formed by crystallization of one magma batch in a closed system and that the lujavrites are sandwiched between the roof series and the kakortokites. This model is illustrated by the zoned pegmatites and the mini-scale model pegmatite described in the present paper which have upper zones rich in sodalite corresponding to the naujaite, lower zones of laminated rocks rich in eudialyte, microcline and arfvedsonite corresponding to the kakortokite, and central parts of aegirine/arfvedsonite-rich rocks which in larger pegmatites are enriched in rare-element minerals and correspond to the lujavrites in the closed system model.

Against the one batch/sandwich model speak the geochemical and mineralogical discontinuities between the roof series rocks and the kakortokite-lujavrite sequence which makes it highly unlikely that these rocks formed from one magma batch. Especially Ba and Sr contents (Table 4), element ratios such as Rb/Sr, Ca/Ba and Zr/La (Table 6), and Nd isotope ratios (Stevenson *et al.* 1997;

Marks *et al.* 2004), display differences between the roof series rocks and the kakortokite-lujavrite sequence that point to a formation by separate magma pulses as proposed by Larsen (1976); Bailey *et al.* (1981) and Sørensen *et al.* (2006).

The mini-scale model pegmatite also provides evidence against the closed system model. The last products of crystallization of this pegmatite, the fine-grained black bodies, which may be correlated with the lujavrite, have very low contents of incompatible elements, in contrast to the elevated contents in lujavrites. This may be referred to exhaustion of supply caused by the copious crystallization of eudialyte in the upper and lower parts of this pegmatite. Similarly, the naujaite and kakortokite are rich in eudialyte. It may be argued that in a closed system, eudialyte crystallization would exhaust the supply of Zr, Hf, REE; Nb, Ta, etc. before the lujavrites crystallized. Furthermore, the magma chamber of the one batch model may not have been large enough to produce the enormous amount of sodalite in the naujaite and cannot explain that fluid inclusions in naujaite sodalite indicate pressures of formation corresponding to depths of 10 km (Markl *et al.* 2001).

The one batch model also meets geometrical problems. The thick naujaite masses in the lowermost part of bore holes II and VI and independent lujavrite occurrences do not fit into the geometrical distribution of rocks above and below the lujavrite horizon (Sørensen *et al.* 2006). Screens of augite syenite occur in the westernmost part of the complex at the Kvanefjeld plateau, in the Narsaq Elv valley, and in Kvanefjeld bore holes to depths of 400 m beneath the surface. Forty of the seventy Kvanefjeld bore holes contain xenoliths of volcanic rocks to depths of 340 m below the surface. These occurrences of augite syenite and volcanic rocks are either xenoliths, which have sunk in the magma, or resistant screens intruded and engulfed by lujavrite and left more or less *in situ* along the north contact of the complex. These left-over screens of augite syenite, naujaite and rocks from the roof of the complex may have divided this part of the magma chamber into a number of sections. No traces of augite syenite and lava xenoliths have been found in bore holes I-VII distributed over the complex. Drill core V contains, however, a number of cm-size bodies of fine-grained black rocks of lujavritic composition but of unknown origin (Rose-Hansen & Sørensen 2002).

If sodalite began crystallization deep in the crust, the intruding naujaitic magma must have spread out to form the more than 600 m thick naujaite zone across the whole complex. The mini-scale model pegmatite described in the present paper forms a cupola on a pegmatite sheet. Sodalite is concentrated in the upper part of the sheet as well as in the upper part of the cupola. By analogy, if the cupola is regarded to be a mini-scale model of the agpaitic magma chamber, it demonstrates that sodalite from a large volume may collect in an appropriate trap.

(7) The roof of the agpaitic magma chamber appears to have a very irregular form. In the Nakkaalaaq-Kvanefjeld area, it seems that there is some kind of an asymmetric dome with its top at the summit of Nakkaalaaq, a nearly vertical north side and gentle slopes towards the west, east and also south towards Tunulliarfik. South of this fjord the slope is towards the north, an indication that sagging of the central part of the complex has occurred as proposed by Ussing (1912).

Also the contact surface between naujaite and the underlying lujavrites is topographically very irregular as it is seen in the north wall of Tunulliarfik (Fig. 23). The structures in the naujaite rafts in lujavrite are parallel to those in the overlying main body of naujaite and in the down-protruding naujaite flanking the lujavrite 'dome'.

The steep dome structure in the Nakkaalaak-Kvanefjeld area may be a primary feature or may have been caused by a push of a lujavritic intrusion, the ascent of which, however, was arrested at a deeper level. The last-named origin is in best agreement with the parallelity of the contacts between the roof series rocks in the Nakkaalaak-Taseq area, which must be a primary feature.

The exposed and inferred unexposed naujaite horizons in aegirine lujavrites and arfvedsonite lujavrites including the breccia zone, which now forms the boundary between naujaite and lujavrite, mark successive boundaries between the naujaitic roof and the ascending kakortokitic-lujavritic magma. The naujaitic roof was gradually disintegrated by the rising lujavritic magma which intruded fractures in the naujaitic roof and worked its way upward step by step enclosing naujaite rafts by piecemeal stoping.

The scarcity of distinct internal intrusive contacts within the main mass of lujavrite and the lack of roof contacts between the possible successive stages of crystallization of the lujavrite sequence may be explained as follows: aegirine lujavrite not only occurs as the lowermost part of the lujavrite sequence but also as layers and xenoliths in the overlying arfvedsonite lujavrites. In the contacts against the naujaite roof and the enclosed naujaite rafts, there is generally a concentration of aegirine lujavrite (Sørensen *et al.* 2006). This suggests that lujavrite crystallization was generally initiated by crystallization of aegirine lujavrite. The lujavrite magma was volatile-rich and volatiles most probably accumulated below the roof of the lujavritic magma chamber. This may have suspended crystallization temporarily and have obliterated possible contacts between successive stages of consolidation. However, eventually arfvedsonite formed instead of aegirine, and arfvedsonite lujavrite replaced the original aegirine lujavrite to various degrees resulting in partly disintegrated xenoliths of aegirine lujavrite in arfvedsonite lujavrite. In addition to such clear cases of remnants of aegirine lujavrite, arfvedsonite lujavrite practically always contains tiny aegirine needles, an indication of a general early crystallization of this mineral in lujavrites.

The accumulation of volatiles and residual elements under the naujaite in the roof of the lujavritic magma chamber was the cause of reactions between naujaite and the lujavritic magma resulting in alteration of the naujaite (Steenfelt & Bohse 1975) and in the formation of various reaction products (Sørensen 1962; Demin 1972).

At Kvanefjeld (Fig. 1) near the roof of the complex, arfvedsonite lujavrite has transgressed the former cover of augite syenite and roof series rocks and is separated from the volcanic country rocks north of the complex by a zone of marginal pegmatite the matrix of which is partly naujaitic, partly massive-textured. The emplacement of the lujavrite was preceded and/or accompanied by shearing and deformation of the volcanic rocks, augite syenite, alkali syenite and anorthosite (Sørensen *et al.* 1969; Nielsen 1967; Nielsen & Steenfelt 1979). Sheets and dykes of lujavrite infiltrate the deformed rocks and contain a mixed association of xenoliths: augite syenite, alkali syenite, pulaskite, foyaite, naujaite and volcanic rocks from the roof lying in a pell-mell. This indicates a more forceful mode of emplacement than inferred for the main lujavrite body.

(8) Xenoliths of country rocks are confined to the contact zones as can be seen at Kvanefjeld and in the west contact and especially in the east contact on the north coast of Tunulliarfik. Such xenoliths, with the exception of augite sye-

nite, do not occur in the interior of the complex. The microsyenite xenolith in the lowermost exposed part of the kakortokite is therefore of considerable interest. It is found at the edge of the best developed trough structure in the whole complex and may have been brought to this position by a magma current. It is silica-oversaturated and cannot be a fragment of a kakortokitic rock. It appears to be related to the microsyenites which occur as dykes in kakortokite and aegirine lujavrite and in the country rocks.

The microsyenites described by Rose-Hansen & Sørensen (2001) are later than the kakortokite which excludes a direct derivation of the xenolith from that suite of rocks. The possibility that the xenolith is a strongly modified basaltic rock from the roof of the complex is highly unlikely. South of Kangerluarsuk, the country rock is the granitic basement which reaches the altitude of 900 m immediately to the west of the complex. Its cover of sandstone and basalt, which has now been eroded away, was separated from the kakortokitic magma chamber by the augite syenite shell and the roof series. Microsyenite dykes are, however, common in the country rocks of the Ilímaussaq complex and are older than the complex (Allaart 1969). There is some uncertainty whether the microsyenite dykes intrude the augite syenite shell or not (Allaart 1969; Larsen & Steenfelt 1974; Marks & Markl 2003). In any case, the most likely source of the xenolith is the regional suite of microsyenites. The xenolith may have survived because it was rapidly shielded by a pegmatitic shell, but it has been mineralogically and chemically modified and is enriched in a number of trace elements. Nielsen & Steenfelt (1979) and Rose-Hansen & Sørensen (2001) concluded that these syenitic intrusions are of external origin and may be related to the regional dyke swarms (Upton *et al.* 2003 and references therein).

(9) The views on the petrological evolution of the Ilímaussaq alkaline complex have changed with time. A certain measure of agreement followed the publication of Larsen & Sørensen (1987).

There are three successive major intrusive stages: augite syenite, alkali granite and nepheline syenite. They all originated in a deep-seated fractionating alkali basaltic magma chamber. It is concluded that there were at least two periods of emplacement of nepheline syenitic melts. That such melts were available is documented by regional dykes of the appropriate compositions (Larsen 1979; Marks & Markl 2003; Sørensen *et al.* 2006).

A first pulaskitic magma pulse formed the roof series which crystallized from the roof downward. The naujaite was formed by abundant crystallization of sodalite which most probably took place throughout the entire magma chamber and perhaps even deeper in the crust during the ascent of the magma from depths. The low density of sodalite made sodalite crystals rise in the magma and collect in a smaller volume as it is demonstrated by the mini-scale model pegmatite. Naujaite formed over a considerable depth interval which is shown by horizons of naujaite rafts in kakortokite, aegirine lujavrite and arfvedsonite lujavrite and by the occurrence of naujaite along the contact between marginal pegmatite and augite syenite at a deep level in the complex. A second magma pulse formed the marginal pegmatite-kakortokite-lujavrite sequence that intruded the roof series from below by piecemeal stoping. Some lujavrites may have been formed by pulses from separate magma pockets.

The irregular shapes of the upper contact surface of the Ilímaussaq magma chamber and of the contact surface between naujaite and the underlying lujavrites may be primary, but more likely originated by a combination of sagging and perhaps roof collapse of the central part of the complex and by piecemeal stoping, and in a late stage perhaps by forceful intrusion.

Outlook

This survey and discussion of the petrology of the Ilímaussaq alkaline complex has demonstrated that, in spite of the dedicated work of a multinational group of researchers, satisfactory up-to-date mapping, description and interpretation are still incomplete. There is room for new observations and new views on the origin, composition, structure and evolution of the complex.

One of the aims of the present paper was to produce information about the magmatic evolution of the complex by examination of rocks, the chemical analyses of which may be equivalent to the chemical composition of the melt from which they formed. Unfortunately, some of the rocks collected for this purpose, when examined in the laboratory, were found to be altered by late- and perhaps post-magmatic processes, a feature they share with other massive-textured Ilímaussaq nepheline syenites. The detailed examination of the samples showed, fortunately, that they nevertheless provide new information about melt compositions and melt evolution. It is, however, evident that more work is needed in order to confirm and constrain the data and the partly new views presented in this paper.

The whole complex was mapped at the 1: 20 000 scale by John Ferguson resulting in a for that time excellent map published in 1964. The southern part of the complex has been remapped resulting in a very detailed new 1 : 20 000 map (Andersen *et al.* 1988). A comparison of Ferguson's map with the map of Andersen *et al.* (1988) clearly shows that it is desirable also to produce a detailed map of the part north of Tunulliarfik. The present paper's description of hitherto unknown occurrences of marginal pegmatite in the west and east contacts of the complex is one of many examples which emphasise the need for a remapping of the northern half of the complex.

The Kvanefjeld area in the northern part of the complex has been mapped on the scale of 1: 2 000 and explored by 70 bore holes with a total core length of 12 km. Additionally, seven bore holes, I-VII, each 200 m deep, are scattered over the complex, and a number of holes penetrate aegirine lujavrite and the uppermost part of the kakortokite in the southern part of the complex. Targets for renewed drilling would be bore holes penetrating the deepest part of the kakortokites on the south coast of Kangerluarsuk, the transition from aegirine lujavrite to kakortokite, the 'dome' structure under Nakkaalaak mountain, the deepest part of the lujavrites near the east contact on the north coast of Tunulliarfik and some bore holes through the marginal pegmatite. Such bore holes would add important new information on the architecture and evolution of the complex and may furthermore supply information of importance for the exploration of the mineral wealth of the complex. Questions to answer are, what is below the kakortokite, is the stratigraphy observed in the exposed part also present in the non-exposed part, are there separate lujavritic magma chambers, did the complex form in a closed or open system, the role and place of the Kvanefjeld alkali syenite, and how many major magma injections were involved in the formation of the complex? Also a more detailed examination of the west and east contacts and of the lujavrites intruding the country rocks is needed. The west point of Nunasarnaq may supply important information about lujavrite stratigraphy and emplacement.

The heat flow measurements of Sass *et al.* (1972) were made in bore holes on the Kvanefjeld plateau at a high level in the complex in rocks with high contents of U, Th and K. As stated by Sass *et al.*

(1972: 6440): 'The large range of radioactivity and the complicated structure of Ilímaussaq present a formidable sampling problem. In particular, the determination of the effective heat production for the intrusion and its relation to heat flow observed at Kvanefjeld is very difficult'. In fact, the bore holes used for the measurement of heat flow were dominated by lujavritic rocks with elevated contents of U, Th and K. Knowledge about the depth to the base of the concealed part of the agpaitic rocks is essential for evaluating which model for the evolution of the agpaitic rocks to choose. New heat flow measurements in bore holes through the less radioactive kakortokite and additional geophysical surveying would certainly contribute to the clarification of this problem.

The partly contradictory results of the rather few isotope system analyses show the need for additional work in this area.

The Ilímaussaq complex is unique with regard to its wealth of rare elements and exotic rocks and minerals, and to the information it provides about magma chamber processes. Measures should be taken to secure that the complex is made available for the international scientific community and interested amateurs and that it is protected against unqualified and destructive activities.

Acknowledgements

The present study is based on field work carried out from 1955 to 2004 supported by the Geological Survey of Greenland, GGU (now the Geological Survey of Denmark and Greenland, GEUS), the Geological Institute (GI) of University of Copenhagen, the Danish Atomic Energy Commission (dissolved in 1978), the Danish Natural Science Research Council (dissolved in 2005) and in the last years by the Carlsberg Foundation.

Through all years, Geological Institute has offered excellent facilities and support, chemical analyses have been produced by GEUS and GI and assistance with regard to preparation of thin sections, maps, diagrams, photos has been offered by generations of staff members at GI, GGU and GEUS. Part of the field and laboratory studies has been carried out in close cooperation with the Danish National Research Laboratory RISØ.

So many colleagues and students from many countries have taken part in the field work, in summer schools and excursions that it is impossible to mention. I extend my sincere thanks to all who participated in the field and laboratory activities but also to the many colleagues with whom I have had the priviledge to discuss the Ilímaussaq complex and agpaitic nepheline syenites at conferences, workshops, field visits and excursions in Ilímaussaq and around the world. No one mentioned, nobody forgotten, all remembered. Lotte Melchior Larsen and John Bailey have kindly read the manuscript and made valuable suggestions for improvements.

References

Allaart, J.H. 1969. The chronology and petrography of the Gardar dykes between Igaliko Fjord and Redekammen, South Greenland. *Rapport Grønlands Geologiske Undersøgelse* 25, 26 pp.

Andersen, S., Bohse, H. & Steenfelt, A., 1981a. A geological section through the southern part of the Ilímaussaq intrusion. *Rapport Grønlands Geologiske Undersøgelse* 103, 39-42.

Andersen, S., Bailey, J.C. & Bohse, H. 1981b. The Zr-Y-U stratigraphy of the kakortokite-lujavrite sequence, southern Ilímaussaq intrusion. *Rapport Grønlands Geologiske Undersøgelse* 103, 69-76.

Andersen, S., Bohse, H. & Steenfelt, A. 1988. *Geological map of Greenland 1:20 000 the southern part of the Ilímaussaq complex, South Greenland*. Grønlands Geologiske Undersøgelse/Geodætisk Institut, Denmark.

Andersen, T. & Sørensen, H. 1993. Crystallization and metasomatism of nepheline syenite xenoliths in quartz-bearing intrusive rocks in the Permian Oslo rift, SE Norway. *Norsk Geologisk Tidsskrift* 73, 250-266.

Andersen, T. & Sørensen, H. 2005. Stability of naujakasite in hyperagpaitic melts, and the petrology of naujakasite lujavrite in the Ilímaussaq alkaline complex, South Greenland. *Mineralogical Magazine* 69, 125-136.

Bailey, J.C. 1995. Cryptorhythmic and macrorhythmic layering in aegirine lujavrite, Ilímaussaq alkaline intrusion, South Greenland. *Bulletin of the Geological Society of Denmark* 42, 1-16.

Bailey, J.C., Larsen, L.M. & Sørensen, H. 1981. Introduction to the Ilímaussaq intrusion with a summary of the reported investigations. *Rapport Grønlands Geologiske Undersøgelse* 103, 5-17.

Bailey, J.C., Gwozdz, R., Rose-Hansen, J. & Sørensen, H. 2001. Geochemical overview of the Ilímaussaq alkaline complex, South Greenland. In: Sørensen, H. (ed). *The Ilímaussaq Alkaline Complex, South Greenland, Status of Mineralogical Research with New Results*. Geology of Greenland Survey Bulletin 190, 35-54.

Bailey, J.C., Sørensen, H., Andersen, T., Kogarko, L.N. & Rose-Hansen, J. 2006. On the origin of microrhythmic layering in arfvedsonite lujavrites from the Ilímaussaq alkaline complex, South Greenland. *Lithos 91*.

Blundell, D.J. 1978. A gravity survey across the Gardar Igneous Province, SW Greenland. *Journal of the Geological Society (London)* 135, 545-554.

Bohse, H. & Andersen, S. 1981. Review of the stratigraphic divisions of the kakortokite and lujavrite in southern Ilímaussaq. *Rapport Grønlands Geologiske Undersøgelse* 103, 53-62.

Bohse, H., Brooks, C.K. & Kunzendorf, H. 1971. Field observations on the kakortokites of the Ilímaussaq intrusion, South Greenland, including mapping and analyses by portable X-ray fluorescence equipment for zirconium and niobium. *Rapport Grønlands Geologiske Undersøgelse* 38, 43 pp.

Bohse, H., Rose-Hansen, J., Sørensen, H., Steenfelt, A., Løvborg, L. & Kunzendorf, H. 1974. On the behaviour of uranium during crystallization of magmas - with special emphasis on alkaline magmas. In: *Formation of Uranium Ore Deposits*. 49-60. Vienna: International Atomic Energy Agency.

Bridgwater, D. & Harry, W.T. 1968. Anorthosite xenoliths and plagioclase megacrysts in Precambrian intrusions of South Greenland. *Bulletin Grønlands Geologiske Undersøgelse* 77, 243 pp. (Also Meddelelser om Grønland 185(2)).

Daly, R. A. 1903. The mechanics of igneous intrusions. American Journal of Science 4th Series 15, 269; 16, 107.

Demin, A. 1972. Interpretation of field observations in the southern part of the Ilímaussaq intrusion, South Greenland. *Rapport Grønlands Geologiske Undersøgelse* 45, 333-335.

Emeleus, C.H. & Harry, W.T. 1970. The Igaliko nepheline syenite complex, general description. *Bulletin Grønlands Geologiske Undersøgelse* 85, 115 pp. (also Meddelelser om Grønland 186(3)).

Emeleus, C.H. & Upton, B.G.J. 1976. The Gardar period in southern Greenland. In: Escher, A. and Watt, W.S. (eds). *Geology of Greenland*, 152-181. Copenhagen: Geological Survey of Greenland.

Engell, J. 1973. A closed system crystal-fractionation model for the agpaitic Ilímaussaq intrusion, South Greenland, with special reference to the lujavrites. *Bulletin of the Geological Society of Denmark* 22, 334-362.

Ferguson, J. 1964. Geology of the Ilímaussaq alkaline intrusion, South Greenland. *Bulletin Grønlands Geologiske Undersøgelse* 39, 82 pp. (also Meddelelser om Grønland 172(4)).

Ferguson, J. 1970a. The significance of the kakortokite in the evolution of the Ilímaussaq intrusion, South Greenland. *Bulletin Grønlands Geologiske Undersøgelse* 89, 193 pp. (also Meddelelser om Grønland 190(1)).

Ferguson, J. 1970b. The differentiation of agpaitic magmas: the Ilímaussaq intrusion, South Greenland. *Canadian Mineralogist* 10, 335-349.

Forsberg, R. & Rasmussen, K. L. 1978. Gravity and rock densities in the Ilímaussaq area, South Greenland. *Rapport Grønlands Geologiske Undersøgelse* 90, 81-84.

Gerasimovsky, V.I. 1969. *Geochemistry of the Ilímaussaq Alkaline Massif (South-West Greenland)*, 174 pp. Moscow: Nauka (in Russian).

Gerasimovsky, V.I., Volkov, V.P., Kogarko, L.N., Polyakov, A.I., Sapryikina, T.V. & Balashov, Ju.A. 1966. *Geochemistry of the Lovozero Alkaline Massif*, 395 pp. Moscow: Nauka (in Russian). English translation by D.A. Brown: Canberra, Australian National University Press, 1968.

Guttenberger, R. von & LeCouteur, P. 1992. Geology and ore reserves. Ilímaussaq project Greenland for Highwood Resources Ltd., Platinova A.S, Calcas A.S. 4 volumes. Unpublished report *Strathcona Mineral Services Ltd. Toronto, Canada.*

Hamilton, E.I. 1964. The geochemistry of the northern part of the Ilímaussaq intrusion, S.W.Greenland. *Bulletin Grønlands Geologiske Undersøgelse* 42, 104 pp. (also Meddelelser om Grønland 162(10).

Jones, A.P. & Larsen, L.M., 1985. Geochemistry and REE minerals of nepheline syenites from the Motzfeldt Centre, South Greenland. *American Mineralogist* 70, 1087-1100.

Johnsen, O., Grice, J.D. & Gault, R.A. 2001. The eudialyte group: a review. In: Sørensen, H. (ed). *The Ilímaussaq Alkaline Complex, South Greenland, Status of Mineralogical Research with new Results*. Geology of Greenland Survey Bulletin 190, 65-72.

Khomyakov, A.P., Sørensen, H., Petersen, O.V. & Bailey, J.C., 2001. Naujakasite from the Ilímaussaq alkaline complex, South Greenland, and the Lovozero alkaline complex, Kola Peninsula, Russia. In: Sørensen, H. (ed). *The Ilímaussaq Alkaline Complex, South Greenland, Status of Mineralogical Research with new Results*. Geology of Greenland Survey Bulletin 190, 95-108.

Kogarko, L.N. 1977. *Genetic Problems of Agpaitic Magmas*. Moscow: Nauka, 294 pp. (in Russian)

Krumrei, T.V., Villa, I.M., Marks, M.A.W. & Markl, G. 2006. A $^{40}Ar/^{39}Ar$ and U/Pb isotopic study of the Ilímaussaq complex, South Greenland: Implications for the ^{40}K decay constant and for the duration of magmatic activity in a peralkaline complex. *Chemical Geology* 227, 258-273.

Kystol, J. & Larsen, L.M. 1999. Analytical procedures in the Rock Geochemical Laboratory of the Geological Survey of Denmark and Greenland. *Geology of Greenland Survey Bulletin* 184, 59-62.

Larsen, J.G. 1977. Petrology of the late lavas of the Eriksfjord Formation, Gardar provincce, South Greenland. *Bulletin Grønlands Geologiske Undersøgelse* 125, 31 pp.

Larsen, L.M., 1976. Clinopyroxenes and coexisting mafic minerals from the alkaline Ilímaussaq intrusion, South Greenland. *Journal of Petrology* 17, 258-290.

Larsen, L.M., 1977. Aenigmatites from the Ilímaussaq intrusion, south Greenland: chemistry and petrological implications. *Lithos* 10, 257-270.

Larsen, L.M. 1979. Distribution of REE and other trace elements between phenocrysts and peralkaline undersaturated magmas, exemplified by rocks from the Gardar igneous province, south Greenland. *Lithos* 12, 303-315.

Larsen, L.M. & Sørensen, H. 1987. The Ilímaussaq intrusion - progressive crystallization and formation of layering in an agpaitic magma. In: Fitton, J.G. & Upton,

B.G.J. (eds). *Alkaline Igneous Rocks*. Special Publication Geological Society (London) 30, 473-488.

Larsen, L.M. & Steenfelt, A. 1974. Alkali loss and retention in an iron-rich peralkaline phonolite dyke from the Gardar province, south Greenland. *Lithos* 7, 81-90.

Macdonald, R. 1974. The rôle of fractional crystallization in the formation of alkaline rocks. In: Sørensen, H. (ed). *The Alkaline Rocks*, 442-459. London: John Wiley & Sons.

Markl, G., Marks, M., Schwinn, G. & Sommer, H. 2001. Phase equilibrium constraints on intensive crystallization parameters of the Ilímaussaq complex, South Greenland. *Journal of Petrology* 42, 2231-2258.

Marks, M. & Markl, G. 2001. Fractionation and assimilation processes in alkaline augite syenite unit of the Ilímaussaq intrusion, South Greenland, as deduced from phase equilibria. *Journal of Petrology* 42, 1947-1969.

Marks, M. & Markl, G. 2003. Ilímaussaq 'en miniature', closed-system fractionation in an agpaitic dyke rock from the Gardar Province, South Greenland. *Mineralogical Magazine* 67, 893-919.

Marks, M., Venneman, T., Siebel, W. & Markl, G. 2004. Nd-, O- and H-isotope evidence for complex, closed system fluid evolution of the peralkaline Ilímaussaq intrusion, South Greenland. *Geochimica et Cosmochimica Acta* 68, 3379-3395.

Nielsen, B.L. 1967. *Anorthosit Forekomster på Kvanefjeldet. Forekomst og Dannelse*. 127 pp. Unpublished cand.scient. thesis, Institut for Petrologi, Københavns Universitet, Danmark.

Nielsen, B.L. & Steenfelt, A., 1979. Intrusive events at Kvanefjeld in the Ilímaussaq igneous complex. *Bulletin of the Geological Society of Denmark* 27, 143-155.

Norrish, K. & Chappel, B.W. 1977. X-ray fluorescence spectrometry. In: Zussman, J. (ed.): *Physical Methods in Determinative Mineralogy*. London: Academic Press 2nd Ed., 201-272.

Petersen, O.V. 2001. List of minerals identified in the Ilímaussaq alkaline complex, South Greenland. In: Sørensen, H. (ed). *The Ilímaussaq Alkaline Complex, South Greenland, Status of Mineralogical Research with New Results*. Geology of Greenland Survey Bulletin 190, 25-33.

Petersen, O.V., Micheelsen, H.I. & Leonardsen, E.S. 1995. Bavenite, $Ca_4Be_3Al[Si_9O_{25}(OH)_3]$, from the Ilímaussaq alkaline complex, South Greenland. *Neues Jahrbuch für Mineralogie Monatshefte* 1995, 321-335.

Ridolfi, F., Renzulli, A., Santi, P. & Upton, B.G.J. 2003. Evolutionary stages of crystallization of weakly peralkaline syenites: evidence from ejecta in the plinian deposits of Agua de Pau volcano (São Miguel, Azores Islands). *Mineralogical Magazine* 67, 749-767.

Rose-Hansen, J. & Sørensen, H. 2001. Minor intrusions of peralkaline microsyenites in the Ilímaussaq alkaline complex, South Greenland. *Bulletin of the Geological Society of Denmark* 48, 9-29.

Rose-Hansen, J. & Sørensen, H. 2002. Geology of lujavrites from the Ilímaussaq alkaline complex, South Greenland with information from seven bore holes. *Greenland Geoscience* 40, 58 pp.

Rose-Hansen, J., Sørensen, H. & Watt, W.S., 2001. Inventory on the literature on the Ilímaussaq alkaline complex, South Greenland. *Danmark og Grønlands Geologiske Undersøgelse Rapport* 2001/102, 42 pp.

Rose-Hansen, J., Karup-Møller, S., Sørensen, E. & Sørensen, H. 1977. Albitites from the northern contact of the Ilímaussaq alkaline intrusion. *Rapport Grønlands Geologiske Undersøgelse* 85, 68 only.

Sass, J.H., Nielsen, B.L., Wollenberg, H.A. & Munroe, R.J. 1972. Heat flow and surface radioactivity at two sites in South Greenland. *Journal of Geophysical Research* 77, 6435-6444.

Semenov, E.I. 1969. *Mineralogy of the Ilímaussaq Alkaline Massif (South Greenland)*, 165 pp. Moscow: Nauka (in Russian).

Sood, M.K. & Edgar, A.D. 1970. Melting relations of undersaturated alkaline rocks from the Ilímaussaq intrusion and Grønnedal-Íka complex, South Greenland, under water vapor and controlled partial oxygen pressure. *Meddelelser om Grønland* 181(12), 41 pp.

Sørensen, H. 1958. The Ilímaussaq batholith. A review and discussion. *Bulletin Grønlands Geologiske Undersøgelse* 19, 48 pp. (also Meddelelser om Grønland 162(3)).

Sørensen, H. 1962. On the occurrence of

steenstrupine in the Ilímaussaq massif, southwest Greenland. *Bulletin Grønlands Geologiske Undersøgelse* 32, 251 pp. (also Meddelelser om Grønland 167(1)).

Sørensen, H. 1966. On the magmatic evolution of the alkaline igneous province of South Greenland. *Rapport Grønlands Geologiske Undersøgelse* 7, 19 pp. Also published in Russian in Problemmyi Geochimii. Moskva: Institute of Geochemistry and Analytical Chemistry 1965, 338-349.

Sørensen, H. 1968. Über die Paragenese einiger alkalischer Gesteine in Südgrönland. *Freiberger Forschungsheft* C 230, 177-186.

Sørensen, H. 1969. Rhythmic igneous layering in peralkaline intrusions. An essay review on Ilímaussaq (Greenland) and Lovozero (Kola, USSR). *Lithos* 2, 261-283.

Sørensen, H. 1970. Internal structures and geological setting of the three agpaitic intrusions - Khibina and Lovozero of the Kola Peninsula and Ilímaussaq, South Greenland. *Canadian Mineralogist* 10, 299-334.

Sørensen, H. 1978. The position of the augite syenite and pulaskite in the Ilímaussaq intrusion, South Greenland. *Bulletin of the Geological Society of Denmark* 27, special issue, 15-23.

Sørensen, H. 2001. Brief introduction to the geology of the Ilímaussaq alkaline complex, South Greenland, and its exploration history. In: Sørensen, H. (ed). *The Ilímaussaq Alkaline Complex, South Greenland, Status of Mineralogical Research with New Results*. Geology of Greenland Survey Bulletin 190, 7-24.

Sørensen, H. 2003. Development of nepheline syenites in rift zones – information from three rift complexes. *Geolines* 15, 140-146.

Sørensen, H. & Larsen, L.M. 1987. Layering in the Ilímaussaq alkaline intrusion, South Greenland. In: Parsons, I. (ed). *Origins of Igneous Layering*. NATO Advanced Science Institute Series C. Mathematical and Physical Sciences 196, Dordrecht: D. Reidel Publishing Company, 1-28.

Sørensen, H. & Larsen, L.M., 2001. The hyperagpaitic stage in the evolution of the Ilímaussaq alkaline complex, South Grenland. In: Sørensen, H. (ed). *The Ilímaussaq Alkaline Complex, South Greenland, Status of Mineralogical Research with New Results*. Geology of Greenland Survey Bulletin 190, 83-94.

Sørensen, H., Bohse, H. & Bailey, J.C. 2006. The origin and mode of emplacement of lujavrites from the Ilímaussaq alkaline complex, South Greeenland. *Lithos 91*.

Sørensen, H., Hansen, J. & Bondesen, E. 1969. Preliminary account of the geology of the Kvanefjeld area of the Ilímaussaq intrusion, South Greenland. *Rapport Grønlands Geologiske Undersøgelse* 18, 1-40.

Sørensen, H., Rose-Hansen, J., Nielsen, B.L., Løvborg, L., Sørensen, E. & Løvborg, T. 1974. The uranium deposit at Kvanefjeld, the Ilímaussaq intrusion, South Greenland. Geology, reserves, beneficiation. *Rapport Grønlands Geologiske Undersøgelse* 60, 54 pp.

Sørensen, H., Bailey, J.C., Kogarko, L.N., Rose-Hansen, J. & Karup-Møller, S. 2003. Spheroidal structures in arfvedsonite lujavrite, Ilímaussaq alkaline complex, South Greenland – an example of macro-scale liquid immiscibility. *Lithos* 70, 1-20.

Steenfelt, A., 1972. *Beskrivelse af Pulaskit, Heterogen Foyait, Sodalitfoyait, Naujait og Kakortokit på Kvanefjeldsplateauet, Ilímaussaq*. 52 pp + appendices. Unpublished cand. scient. thesis, Geologisk Institut, København.

Steenfelt, A., 1981. Field relations in the roof zone of the Ilímaussaq intrusion with special reference to the alkali-acid rocks. *Rapport Grønlands Geologiske Undersøgelse* 103, 43-52.

Steenfelt, A. & Bohse, H. 1975. Variations in the content of uranium in eudialyte from the differentiated alkaline Ilímaussaq intrusion, south Greenland. *Lithos* 8, 39-45.

Stephenson, D. 1976. A simple-shear model for the ductile deformation of high-level intrusions in South Greeenland. *Journal of the Geological Society (London)* 132, 307-318.

Stephenson, D. & Upton, B.G.J. 1982. Ferromagnesian silicates from a differentiated alkaline complex: Kûngnât Fjeld, South Greenland. *Mineralogical Magazine* 46, 283-300.

Stevenson, R., Upton, B.G.J. & Steenfelt, A. 1997. Crust-mantle interaction in the evolution of the Ilímaussaq Complex, South Greenland: Nd isotopic studies. *Lithos* 40, 189-202.

Upton, B.G.J., 1996. Anorthosites and tractolites of the Gardar Province. In: Demaiffe, D. (ed). *Petrology and Geochemistry of Magmatic Suites of Rocks in Continental and Oceanic Crusts*. A volume dedicated to Professor Jean Michot: Université Libre de Bruxelles, Royal Museum for Central Africa (Tevuren) 19-34.

Upton, B.G.J. & Blundell, D.J. 1978. The Gardar igneous province: evidence for Proterozoic continental rifting. In: Neumann, E.-R. & Ramberg, I.B. (eds). *Petrology and Geochemistry of Continental Rifts*. NATO Advanced Science Institutes, Series C. Matematical & Physical Sciences 163-172. Dordrecht: Reidel Publishing Company.

Upton, B.G.J. & Emeleus, C.H. 1987. Mid-Proterozoic alkaline magmatism in southern Greenland: the Gardar province. In: Fitton, J.G. and Upton, B.G.J. (eds). *Alkaline Igneous Rocks*. Geological Society Special Publication (London) 30, 449-471.

Upton, B.G.J., Stephenson, D. & Martin, A.R. 1985. The Tugtutôq Older Giant Dyke Complex: mineralogy and geochemistry of an alkali gabbro – augite syenite - foyaite association in the Gardar Province of South Greenland. *Mineralogical Magazine* 49, 623-642.

Upton, B.G.J., Emeleus, C.H., Heaman, L.M., Goodenough, K.M. & Finch, A.A. 2003. Magmatism of the mid-Proterozoic Gardar Province, South Greenland: chronology, petrogenesis and geological setting. *Lithos* 68, 43-65.

Ussing, N.V., 1894. Mineralogisk-petrografisk Undersøgelse af Grønlandske Nefelinsyeniter og beslægtede Bjærgarter. *Meddelelser om Grønland* 14, 120 pp.

Ussing, N.V. 1912. Geology of the country around Julianehaab, Greenland. *Meddelelser om Grønland* 38, 426 pp.

Waight, T., Baker, J. & Willigers, B. 2002. Rb isotope dilution analyses by MC-ICPMS using Zr to correct for mass fractionation: towards improved Rb-Sr geochronology? *Chemical Geology* 186, 99-116.

Wolff, J.A. & Toney, J.B. 1993. Trapped liquid from a nepheline syenite: a re-evaluation of Na-, Zr-, F-rich interstitial glass in a xenolith from Tenerife, Canary Islands. Lithos 29, 285-293.